# Thermal Data
for
# Natural and
# Synthetic Fuels

# Thermal Data
## for
# Natural and
# Synthetic Fuels

## SIDDHARTHA GAUR
*VSLR Sciences*
*Dallas, Texas*

## THOMAS B. REED
*Colorado School of Mines*
*Golden, Colorado*

MARCEL DEKKER, INC.          NEW YORK · BASEL · HONG KONG

Library of Congress Cataloging-in-Publication Data

Gaur, Siddhartha.
    Thermal data for natural and synthetic fuels / Siddhartha Gaur,
Thomas B. Reed.
        p.   cm.
    Includes bibliographical references and index.
    ISBN 0-8247-0070-8  (alk. paper)
    1. Fuel--Thermal properties.   2. Synthetic Fuels--Thermal
properties.   I. Reed, Thomas B.   II. Title.
TP321.G37   1998
662'.6--dc21

                                                                    98-16718
                                                                        CIP

This book is printed on acid-free paper.

**Headquarters**
Marcel Dekker, Inc.
270 Madison Avenue, New York, NY 10016
tel: 212-696-9000; fax: 212-685-4540

**Eastern Hemisphere Distribution**
Marcel Dekker AG
Hutgasse 4, Postfach 812, CH-4001 Basel, Switzerland
tel: 44-61-261-8482; fax: 44-61-261-8896

**World Wide Web**
http://www.dekker.com

The publisher offers discounts on this book when ordered in bulk quantities. For more information, write to Special Sales/Professional Marketing at the headquarters address above.

Current printing (last digit):
10  9  8  7  6  5  4  3  2  1

**PRINTED IN THE UNITED STATES OF AMERICA**

To Poorva, Sierra, and Mansi

# PREFACE

Thermal data such as proximate and ultimate analysis have been used routinely in evaluating coal, biomass, and other fuels for over a century. In the last few decades, additional thermal analysis (TA) tools such as thermogravimetry, differential thermal analysis, and thermal mechanical analysis have received significant attention in the evaluation of thermal behavior of these substances. The data obtained from these techniques can provide useful information in terms of reaction mechanism, kinetic parameters, thermal stability, phase transformation, heat of reaction, and other aspects of gas–solid and gas–liquid systems. Unfortunately, there are no ASTM standards set for the use of TA techniques, and therefore investigators use conditions that suit their requirements for measuring the relevant thermal data. Because of this lack of standardization, the information obtained from two different sources is not comparable.

Thermal analysis data are useful for both research and engineering applications. In the case of research scientists, identification of different reaction mechanisms, determination of kinetic parameters, and optimization of conditions to favor one reaction over the other are some of the necessary information that can be obtained from thermal analysis data. In the case of engineering applications, thermal analysis data on specific temperatures at which various heterogeneous reactions occur, their reaction rates, and the energies involved in these reactions are invaluable information for process designing.

This book presents thermogravimetric (TG) and differential thermal analysis (DTA) data gathered under comparable conditions on a wide variety of organic materials such as biomass, coal, municipal solid wastes, and some liquid fuels, making it easy to find and compare data on this large variety of samples. Some data have been collected by changing the TG operating conditions to show the effects of such parameters as particle size and heating rate on the decomposition pattern of a solid sample. This information can be used to provide the linkage and guideline for the extrapolation of the previous data available in literature but obtained at different conditions.

*Thermogravimetric analysis* records the weight change of a sample as a function of time or temperature in a preset time–temperature program. A typical thermogram illustrating weight change as a function of both time and temperature as obtained from Seiko TG/DTA SSC 5000 is shown in Figure 1 for a Western red cedar sample heated at 10°C/min in a flowing nitrogen atmosphere. Figure 2 gives the same information on the temperature scale in the format used to present all the data in this book. The temperature at which there is the first sign of weight loss in the sample is shown as $T_{initial}$. $T_{start\text{-}point}$ is shown as the beginning of the inverse S curve as obtained from the method for determining glass transition temperature, and $T_{end\text{-}point}$ is the temperature at the end of the curve. $T_{mid\text{-}point}$ refers to the mid-point between the beginning and the end of the inverse S curve.

It is useful to present the nonisothermal reaction rate data with respect to temperature in order to determine the reaction kinetic parameters that are valid over a wide range of temperature intervals. This can also be done by employing the technique of *differential thermogravimetric analysis* (DTG) as shown in Figure 2. DTG helps in providing the information of kinetic parameters, as well as in distinguishing between the various reaction steps taking place over the entire temperature. The point $T_{max}$ on this curve shows the temperature at which the maximum rate of reaction occurs.

Differential thermal analysis records the temperature difference and therefore the heat of reaction between the sample and a standard inert material as a function of temperature, thus showing the occurrence of either endothermic or exothermic reaction or phase change. The first inflection point signifies the beginning of the transition and the second inflection point shows the end. The

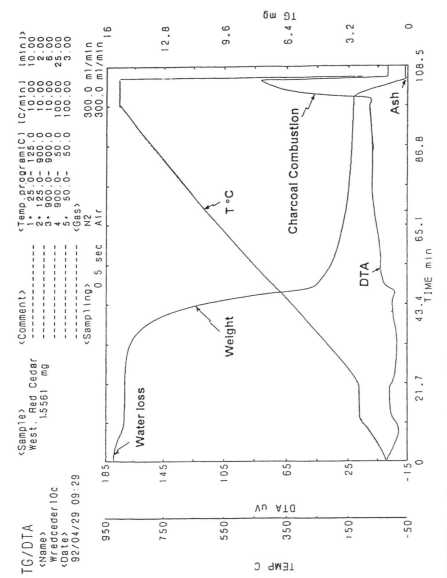

**FIGURE 1** Typical experimental data on Western red cedar sample as obtained from Seiko TG/DTA, showing heating program, temperature curve (T), weight change curve (TG), differential thermal change curve (DTA), and char combustion in air as a function of time. The final weight remaining is equal to total inorganic ash remaining in the sample.

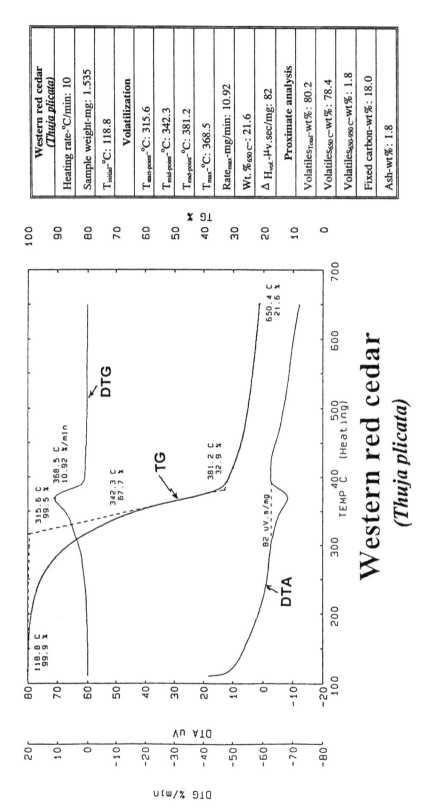

**FIGURE 2** Illustration of data presentation format of this book for thermogravimetric analysis (TG), differential thermogravimetric analysis (DTG), and energy changes during the reaction (DTA). The sidebar shows critical temperatures and proximate analysis.

area within the peak gives an estimate for heat of reaction for that particular transition. The DTA curve for a Western red cedar sample is shown at the bottom of Figure 2.

Another widely used tool in the thermal analysis of organic samples is *proximate analysis* (ASTM D3172-75), which conveys the moisture content of the sample, the total volatile content, the ash content, and an estimate of fixed carbon content. These properties are listed for most samples in the tables accompanying each thermogram. However, the moisture content has been deliberately left out because it is the function of geographic location and ambient humidity. The data presented in the table are given on a moisture-free basis so that they can be compared with the results obtained from any part of the world.

Almost all the data presented in this book are on the same scale coordinates so that the data for different samples can be easily compared. For this purpose, a blank chart is provided (Fig. 3). We suggest that the reader create a transparency to trace different curves and compare them with one another.

Thermogravimetric data are used primarily for the determination of reaction kinetic data and parameter estimation for gas–solid reactions under isothermal and nonisothermal conditions. Keeping this important aspect in view, we have devoted the first three chapters of the book to relevant kinetic theories, an overview of thermal analysis techniques applicable to this kinetic parameter estimation, and the derivation of kinetic data from TG-DTA data.

Chapter 4 provides information about the specific Seiko TG-DTA instrument used to collect all the data presented in this book.

Chapters 5–12 present thermograms of various categories of samples. *Biomass* is the generic name for all natural materials when considered as a source of energy and chemicals. Although biomass occurs in innumerable forms, the principal components are cellulose, hemicellulose, lignin, and sugars. Chapter 5 contains data on these components of biomass.

The thermal data charts are provided in this book to give insight into the thermal behavior of solid fuels when subjected to pyrolytic conditions. However, since this book is intended to be used by researchers and scientists who would also be interested in the thermal behavior of individual components of the organic mass, we have arranged the thermograms by individual biomass components to provide insight into the component behavior. In our opinion, this will help the reader in the analysis of thermal behavior for overall biomass samples obtained from various sources. One reason for placing this chapter early in the book is that there is a school of thought that the overall thermal degradation pattern of biomass

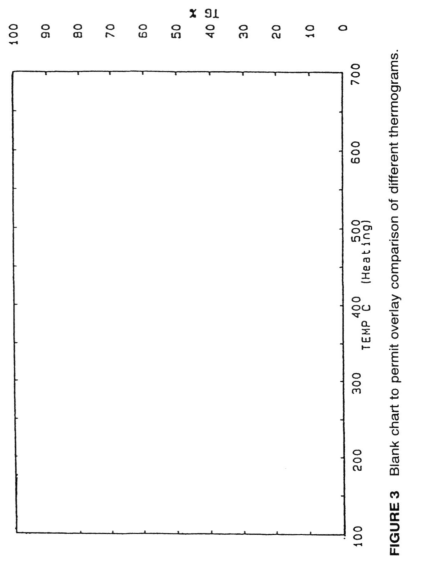

**FIGURE 3**   Blank chart to permit overlay comparison of different thermograms.

samples is the sum of the degradation pattern of its primary components present in the sample. This hypothesis relies on the assumption that the components such as cellulose, hemicellulose, and lignin are noninteractive and that the overall degradation pattern of a biomass sample is proportional to the sum of these components. However, it is important to note that this school of thought is not universally accepted. The presence of thermograms and this chapter at the beginning of the book should aid in the evaluation of this hypothesis as the reader examines different types of biomass samples discussed in subsequent chapters.

Chapter 6 contains TA data on the natural forms of woody, agricultural, and aquatic biomass; Chapter 7 contains information on variations in these thermograms due to plant age, growth conditions, and anatomy. Chapters 8 and 9 present thermal analysis data on processed biomass, such as paper and pellets, and its municipal solid waste (MSW) components, such as plastics. Thermograms on solid fuels such as coal and lignite, which are representative of fuels with high fixed carbon content, are arranged in Chapter 10, while those on liquid fuels are brought together in Chapter 11.

For the sake of uniformity, one set of standard conditions was used for most of the samples presented in this book. By using this approach, TG thermograms can be used as fingerprints for the identification of samples. The specifics of these standards are discussed in Chapter 2.

Nonetheless, it must be noted that setting such standards does raise issues with respect to mass and heat transfer effects. By using small samples and low fixed heating rates, we have minimized these effects to a level where deviations due to them are negligible. However, in doing so we have left a data gap for those researchers interested in understanding the quantitative deviations caused by mass and heat transfer effects. To make up for this omission, Chapter 12 explores thermograms for different sample sizes at different heating rates. These data, along with the kinetic parameters discussed in Chapter 3, will enable the researcher to extrapolate the graphs presented in this book to other conditions of interest.

We present this book on thermal data in the hope that it will be useful for scientists who wish to understand more about the mechanism of pyrolysis and for engineers working in the field who need accurate data to design experiments and projects.

During the preparation of this book, several people provided help in many ways. We wish to thank them all for their contributions: Drs. Thomas Milne, Robert Evans, and Foster Agblevor from the National Renewable Energy Laboratory, for participating throughout the program; Professor Robert M. Baldwin, Head of the

Chemical Engineering and Petroleum Refining Department, and Professor Kent Vorhees, Chemistry Department, for allowing the use of their facilities; Professor Herb Schroder, Colorado State University, Paper Research Institute, Marine Research Institute, and others for the supply of samples; and Mr. James Adams for conducting some of the test runs. We also wish to acknowledge the efforts of Marcel Dekker, Inc.—in particular, Dawn Wechsler, Linda Schonberg, and Steven Sidore—to bring this book to its present format. Finally, we would like to thank our wives, Vibha Bansal and Vivian Reed, and Dr. Gaur's parents for being supportive to us during this work.

*Siddhartha Gaur*
*Thomas B. Reed*

# CONTENTS

*xiii*

# Thermal Data
## for
# Natural and
# Synthetic Fuels

# 1

## SOME ASPECTS OF GENERAL CHEMICAL KINETIC THEORIES USED IN THERMAL ANALYSIS

Chemical kinetics, or reaction kinetics, is the quantitative study of the rate at which chemical reactions occur. This rate can vary from very large values to essentially zero. However, most industrially important reactions occur at rates between these extremes. This rate is influenced by variables such as temperature, pressure, composition of reacting mixtures, and presence of foreign particles.

The rate of reaction, which is also called the speed of the transformation process, for a chemical system can be expressed in terms of the concentration of any one of the reacting substances or of the products that are formed.

If the two materials are brought together and then placed under certain conditions, they transform to form a new substance; we then say that a chemical reaction has occurred. The reasons and the mechanism for this transformation are studied, in general, under chemistry. However, the factors that influence the rate of this transformation are studied under chemical kinetics. This kinetics is essential for the design and sizing of reactors that are used in commercial processes to enable a desired chemical transformation.

Conventional kinetic measurement requires multiple measurements of reaction processes as a function of time at given set of conditions and then these data are used to extrapolate the overall kinetics of a conversion process. This approach works very well for homogeneous reactions. However, heterogeneous reaction factors such as sample heterogeneity and uniform temperature distribution require series of experimentation to obtain accurate and reliable data. In modern day experimental procedures, the kinetics of most heterogeneous reactions are being studied by using thermal analysis techniques like thermogravimetry and differential thermal analysis. These techniques provide accurate kinetic data and also save overall time for experimentation. Since most of the general kinetic theories are also applicable to thermal analysis techniques, in this chapter we have reviewed some of these theories for ready reference.

## 1.1  THERMODYNAMICS AND KINETICS

Thermodynamics provides two important pieces of information needed in reactor design: the maximum possible extent of reaction to which a chemical reaction can proceed at given operating conditions and the heat liberated or absorbed during the reaction. Since these values do not account for any change with respect to time, the values predicted by a thermodynamic calculation are termed equilibrium properties.

Since any process will attain to its equilibrium/maximum value of conversion in infinite time, the information about equilibrium conversion provided by the thermodynamic calculation does not help in reasonable sizing of the reactor. The time available for carrying out a chemical reaction commercially is limited if the process is to be economically feasible. On the other hand, a knowledge of equilibrium conversion is important as a goal with which the actual performance of the reaction equipment can be compared. Without the prior knowledge of the equilibrium conversion, erroneous conclusions might be drawn as to the final yield for a chemical process.

A knowledge of the equilibrium conversion is important in other ways as well. Under certain circumstances (high concentrations of products and low equilibrium constant), the rate of reaction in the reverse direction may be significant. In these cases, the net rate of reaction in the forward direction is equal to the rate in the forward direction minus the rate in the reverse direction. This net rate can also be formulated solely in terms of the rate in the forward direction, provided the equilibrium constant is known. In other words, it is not necessary to make separate experimental determinations of the forward and reverse rates of reaction.

The equilibrium constant $K$ is defined in terms of the equilibrium activities $a_i$ of the reactants and products. For a general reaction

$$aA + bB = cC + dD \qquad (1.1)$$

the equilibrium constant is

$$K = \frac{a_C^c a_D^d}{a_A^a a_B^b} \qquad (1.2)$$

The activities refer to equilibrium conditions in the reaction mixture and are defined as the ratio of the fugacity of the reactant and product in the equilibrium mixture to that in the standard state; i.e.

$$a = \frac{f}{f^0} \qquad (1.3)$$

The other important aspect in reactor design involving the thermodynamics viewpoint relates to overall energy changes. The energy absorbed or evolved during a reaction is due to the differences in structure of the products and the reactants. The exact amount of net energy associated with any reaction depends on the temperatures of the reactants and the products. Since the standard heat of reaction $\Delta H_R$ depends on temperature, it is customary to define a standard heat of reaction, based on 25°C and 1 atm pressure. With this basis, the standard heat of reaction $\Delta H_R$ can be formally defined as the change in enthalpy starting with reactants at 25°C and 1 atm pressure and ending with products at 25°C and 1 atm pressure. If $\Delta H_R$ is negative, heat is evolved and the reaction is *exothermic*. If $\Delta H_R$ is positive, heat is absorbed and the reaction is *endothermic*. This heat of reaction can be evaluated from the heats of formation of the reactants and the products:

$$\Delta H_R = \sum_P \Delta H_f - \sum_R \Delta H_f \qquad (1.4)$$

where the first term on the right side refers to the summation for the products and the second to that of the reactants.

## 1.2  HOMOGENEOUS AND HETEROGENEOUS REACTIONS

In homogeneous reactions, all reacting materials are found within a single phase, be it gas, liquid, or solid. In addition, if the reaction is catalytic, the catalyst must also be in the same phase. For homogeneous systems, the rate of reaction of any reaction component $A$ is defined as

$$r_A = \frac{1}{V}\left(\frac{dN_A}{dt}\right)_{\text{by reaction}} = \frac{\text{(moles of } A \text{ that appear by reaction)}}{\text{(unit volume)(unit time)}}$$

$$(1.5)$$

By this definition, if $A$ is a reaction product, the rate is positive; if it is a reactant, the rate is negative; thus $-r_A$ is the rate of disappearance of reactant $A$. Since the rate of reaction depends on variables such as temperature, pressure, and composition of reacting mixtures, we may write the following for the rate of reaction of component $A$:

$$r_A = f(\text{state of the system})$$
$$= f(\text{temperature, pressure, composition})$$

Since pressure can be determined given the temperature and composition, we may write

$$r_A = f(\text{temperature, composition})$$

In the case of heterogeneous reactions, the reactants and the products are not in a single phase. The kinetics of heterogeneous reactions involves additional reactors like the surface area in case of solids.

## 1.3 THE ARRHENIUS RATE LAW: TEMPERATURE DEPENDENCY OF THE REACTION RATE EXPRESSION

It has been known for many years that increasing the temperature frequently causes a marked increase in the rates of reactions; a useful rough generalization is that the rate is doubled by a rise in temperature of 10°C. In 1887, van't Hoff, arguing on the basis of the variation of the equilibrium constant with the temperature, pointed out that the logarithm of the specific rate of a reaction should be a linear function of the reciprocal of the absolute temperature:

$$\ln K = B - \frac{A'}{T} \tag{1.6}$$

or

$$\frac{d\ln K}{dT} = \frac{A'}{T^2} = \frac{\Delta H}{RT^2} \tag{1.7}$$

Since the equilibrium constant is equal to the ratio of $k$ and $k'$, the forward and reverse rate constants, and the overall heat of the reaction is broken into an energy change for each direction, the above equation can be written as follows:

$$\frac{d \ln k}{dT} - \frac{d \ln k'}{dT} = \frac{E - E'}{RT^2} \tag{1.8}$$

Arrhenius extended this idea for two separate expressions, one for forward and one for the reverse, and successfully applied it to the data relative to a number of reactions; on account of his work the law is usually referred to as the Arrhenius rate law, and may be expressed as

$$k = Ae^{-E/RT} \tag{1.9}$$

where $A$ is a constant usually known as the *frequency factor* of the reaction and $E$ is the *energy of activation*. Arrhenius suggested that in order for the reaction to occur, the reactant molecules must become activated and that there exists an equilibrium between normal (inert) and activated molecules. The energy $E$ represents the energy the molecules have to acquire in order to be capable of undergoing reaction, and accounts for the high temperature coefficient of the rate since a rise in temperature increases markedly the proportion of active molecules. The overall reaction then could be looked upon as taking place by the following two-step process:

Reactants → activated reactants → products

## 1.4  TEMPERATURE DEPENDENCY OF THE FREQUENCY FACTOR AND ACTIVATION ENERGY

The temperature dependency of a reaction as per the Arrhenius rate law can be more appropriately referred to as the temperature dependency of its two terms, namely, the frequency factor and the activation energy. To describe these two terms and their dependency on temperature, several theories have been proposed; collision theory and the activated complex theory are the most common of these.

### 1.4.1  Collision Theory

The Arrhenius theory requires that before a reaction can occur, the molecules of reactants must have an energy of activation $E$ (Figure 1.1) above their normal or average energy. According to the collision theory (Lewis, 1918), the number of product molecules formed per unit time per unit volume is equal to the number of collisions multiplied by a factor $f$. In this theory, molecules are treated as hard spheres. This factor takes into account the fact that only a fraction of collisions involve molecules that possess the necessary excess energy (activation energy).

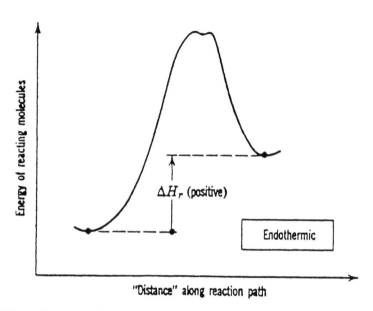

**FIGURE 1.1** Energy of activation for endothermic reactions.

For a simple gaseous reaction such as $A + B \rightarrow C + D$, this may be stated mathematically as follows:

$$r = zf \tag{1.10}$$

where $z$ = number of collisions between molecules per second of 1 cc of reaction mixture. From the kinetic theory for ideal gases, the rate expression can be written as

$$r = f c_A c_B \sigma_{AB}^2 \left( 8\pi RT \, \frac{M_A + M_B}{M_A M_B} \right)^{1/2} \tag{1.11}$$

Using the Arrhenius equations, one can write

$$Ae^{-E/RT} = f\sigma_{AB}^2 \left( 8\pi RT \, \frac{M_A + M_B}{M_A M_B} \right)^{1/2} \tag{1.12}$$

Assuming a Maxwellian distribution, the fraction of the total molecules having an energy $E$ can be shown to be $e^{-E/RT}$. Thus the frequency $A$ is given by

$$A = \sigma_{AB}^2 \left( 8\pi RT \, \frac{M_A + M_B}{M_A M_B} \right)^{1/2} \tag{1.13}$$

and the specific reaction rate is given as

$$k = \sigma_{AB}^2 \left( 8\pi RT \, \frac{M_A + M_B}{M_A M_B} \right)^{1/2} e^{-E/RT} \tag{1.14}$$

Since $T$ is the only variable, one can write

$$A = f(T^{1/2}) \qquad (1.15)$$

This equation is valid for bimolecular reactions. In the case of other types of reactions, the exponential term can be modified accordingly.

## 1.4.2 Activated Complex Theory

The essential feature of this theory is the postulation of an activated complex, an intermediate unstable substance formed from the reactants and decomposing into the products. Thus

$$A + B \rightarrow AB \rightarrow C + D \qquad (1.16)$$

A basic assumption that is made regarding this intermediate activated complex is that it is in thermodynamic equilibrium with the reactants, even when the reactants and products are not at equilibrium with each other. This means that the rate-controlling step in the overall reaction is the rate of decomposition of the activated complex into the products. This concept of an equilibrium activation step followed by a slow decomposition is equivalent to assuming a time lag between activation and decomposition into the products of reaction and can be expressed graphically as depicted in Figure 1.2.

After applying the concepts of quantum mechanics, Eyring, Polanyi and their coworkers showed that the rate constant $k$ can be written as

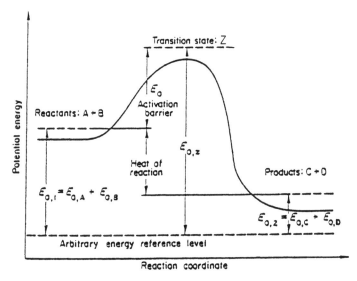

**FIGURE 1.2** Reaction path in terms of reaction coordinate and potential.

$$k = \frac{k_B T}{h} e^{\Delta S/R} e^{-\Delta H/RT} \tag{1.17a}$$

or

$$k = \frac{k_B T}{h} \frac{Q_{AB}}{Q_A Q_B} e^{-\Delta E_0/RT} \tag{1.17b}$$

where $h$ is Planck's constant, $k_B$ is Boltzmann's constant, $\Delta S$ and $\Delta H$ are the entropy and enthalpy of activation, $Q$ refers to the partition function, and $\Delta E_0$ is the energy of activated complex minus that of the reactants. This is the activation energy of the overall reaction at absolute zero. The partition functions in this expression must be evaluated with respect to the zero-point levels of the respective molecules. These functions can be factorized into contributions corresponding to translational, rotational, vibrational, and electronic energy. Table 1.1 gives the corresponding partition functions.

In their original paper, Eyring and Polanyi (1931) included the dynamical treatment of the $H + H_2$ reaction to treat the dynamics of classical motion over potential energy surfaces. The expressions given above show that the frequency factor $A$ is not necessarily independent of temperature. Thus, the rate constant for a reaction can be expressed in the form

$$k = aT^n e^{-E_0/RT} \tag{1.18}$$

The value of the exponent $n$ depends upon the form taken by the partition functions. The fact that plots of the logarithm of $k$ against the reciprocal of the temperature are usually linear for elementary reactions is due to the much stronger temperature dependence of the exponential part than of the pre-exponential part. The experimental energy of activation $E_{\exp}$ is defined by

$$\frac{d \ln k}{dT} = \frac{E_{\exp}}{RT^2} \tag{1.19}$$

since the experimental activation energy is determined by plotting the logarithm of $k$ against the reciprocal of the absolute temperature. Differentiation of the logarithmic form of general expression of rate constant gives

$$\frac{d \ln k}{dT} = \frac{n}{T} + \frac{E_0}{RT^2} = \frac{E_0 + nRT}{RT^2} \tag{1.20}$$

Comparison of these two equations leads to the relationship

$$E_{\exp} = E_0 + nRT \tag{1.21}$$

At the absolute zero, the energy of activation is the difference be-

**TABLE 1.1**

Partition Functions

| Motion | Degrees of freedom | Partition function | Order of magnitude |
|---|---|---|---|
| Translational | 3 | $\dfrac{(2\pi mkT)^{3/2}}{h^2}$ | $10^{24}$–$10^{25}$ |
| Rotational linear molecule | 2 | $\dfrac{8\pi^2 IkT}{h^2}$ | $10$–$10^2$ |
| Rotational nonlinear molecule | 3 | $\dfrac{8\pi^2(8\pi^3 ABC)^{1/2}(kT)^{3/2}}{h^2}$ | $10^2$–$10^3$ |
| Vibrational (per normal mode) | 1 | $\dfrac{1}{1 - e^{-h\nu/kT}}$ | $1$–$10$ |
| Restricted rotation | 1 | $\dfrac{(8\pi^3 I'kT)^2}{h}$ | $1$–$10$ |

where

$m$ = mass of molecule
$I$ = moment of inertia for linear molecule
$A$, $B$, and $C$ = moments of inertia for a nonlinear molecule about three axes at right angles to one another
$I'$ = moment of inertia for restricted rotation
$\nu$ = normal-mode vibrational frequency
$k$ = Boltzmann constant
$h$ = Plank's constant
$T$ = temperature, °K

tween the zero-point levels in the initial and the activated states, whereas at any other temperature it is the difference between the average energies of the reactants and of the activated complex. Unless $n$ is very large or $E_0$ is very small, it may be difficult experimentally to detect this temperature dependence of the energy of activation. Except for trimolecular reactions, this has rarely been done for gas reactions. An example of a trimolecular reaction is $2NO + Cl_2 \rightarrow 2NOCl$. The rate expression for this reaction is of the form

$$k = k'T^{-7/2}e^{-E_0/RT} \tag{1.22}$$

## 1.5  COMPOSITION DEPENDENCY OF THE RATE EXPRESSION

The manner in which the rate of a reaction varies with the concentrations of the reacting substances can be indicated by stating what is called the order of the reaction. Thus a reaction whose rate depends upon the first power of the concentration of one reactant is said to be of the first order, while if the rate is proportional to the product of two reactant concentrations it is of the second order. In general, if it is found experimentally that the rate of a reaction is proportional to the $\alpha$th power of the concentration of one of the reactants $A$, to the $\beta$th power of the concentration of $B$, etc.

$$\text{Rate} = kc_A^\alpha c_B^\beta \dots \tag{1.23}$$

the overall order of the reaction is simply

$$n = \alpha + \beta + \dots \tag{1.24}$$

Such a reaction would be said to be of the $\alpha$th order with respect to $A$, the $\beta$th order with respect to $B$, etc.

Molecularity, on the other hand, refers to the number of molecules involved in the reaction. Thus, for a general reaction

$$aA + bB = cC + dD \tag{1.25}$$

the molecularity is simply

$$m = a + b \tag{1.26}$$

For reactions where the rate expression corresponds to the stoichiometric equation are called elementary reactions. Thus for the above reaction to be elementary, the rate expression must be

$$-r_A = kc_A^a c_B^b \tag{1.27}$$

Also, for such reactions, the molecularity is equal to the overall order of the reaction.

When there is no correspondence between stoichiometry and rate, then we have a nonelementary reaction. The classical example of a nonelementary reaction is that between hydrogen and bromine

$$H_2 + Br \rightarrow 2HBr$$

which has a rate expression

$$r_{HBr} = \frac{k_1 c_{H_2} c_{Br_2}^{1/2}}{k_2 + c_{HBr}/c_{Br_2}} \tag{1.28}$$

Nonelementary reactions are explained by assuming that what we observe as a single reaction is in reality the overall effect of a sequence of elementary reactions. The reason for observing only a single reaction rather than two or more elementary reactions is that the amount of intermediates formed is negligibly small, and therefore escapes detection.

To explain the kinetics of nonelementary reactions, it is assumed that a sequence of elementary reactions are occurring but the intermediates formed are in minute quantities to escape detection. The types of intermediates are mostly suggested by the chemistry of the materials. These may be grouped as follows:

*Free radicals*: Free atoms or larger fragments of stable molecules that contain one or more unpaired electrons are called free radicals. Some free radicals are relatively stable but as a rule they are unstable and highly reactive.

*Ions and polar substances*: Electrically charged atoms, molecules, or fragments of molecules are called ions. These may act as active intermediates in reactions.

*Molecules*: Molecules can act as reactive intermediates.

*Transition complexes*: Collisions between reactant molecules can result in strained bonds, unstable forms of molecules, or unstable association of molecules, which can then decompose to give the product. Such unstable forms are called transition complexes.

## 1.5.1  Types of Reactions

*Single Reaction*

When a single stoichiometric equation and single rate equation are chosen to represent the progress of the reaction, we have a single reaction.

*Multiple Reaction*

When more than one stoichiometric equation is used to represent the observed changes, then more than one kinetic expression is needed to follow the changing composition of all the reaction components, and we may have multiple reactions. Such reactions may be classified as series reactions:

$$A \rightarrow R \rightarrow S \tag{1.29}$$

Parallel reactions are of two types

$$
\begin{array}{ccc}
A \rightarrow R & & A \rightarrow R \\
& \text{and} & \\
A \rightarrow S & & B \rightarrow S
\end{array} \tag{1.30}
$$

competitive     side by side

*Autocatalytic Reaction*

A reaction where one of the products of reaction acts as a catalyst is called an autocatalytic reaction:

$$A + R \rightarrow R + R \tag{1.31}$$

## 1.6 IDENTIFICATION OF
## RATE-CONTROLLING STEPS

Since chemical reaction rates vary so widely from reaction to reaction, and with temperature, very often it is found that one or the other of the two steps provides the major resistance to the overall change. In such a case, this slow step is the rate-controlling step. For example, in solid-catalyzed reactions, various steps are involved that may cause resistances.

*Gas film resistance*: Reactants diffuse from the main body of the fluid to the exterior surface of the catalyst.

*Pore diffusion resistance*: Because the interior of the pellet contains so much more area than the interior, most of the reaction takes place within the particle itself. Therefore, reactants move through the pores into the pellets.

*Surface phenomenon resistance*: At some point in their wanderings, reactant molecules become associated with the surface of the catalyst. They react to give products, which are then released to the fluid phase within the pores.

*Pore diffusion resistance for products*: Products then diffuse out of the pellet.

*Gas film resistance for products*: Products then move from the mouth of catalyst pores into the main gas stream.

*Resistance to gas flow*: For fast reactions accompanied by large heat release or absorption, the flow of heat into or out of the reaction zone may not be fast enough to keep the pellet isothermal. If this happens, the pellet will cool or heat, strongly affecting the rate. Thus the heat transfer across the gas film or within the pellet could influence the rate of reaction.

## 1.7 NONISOTHERMAL REACTION ANALYSIS

Since the first investigation of the kinetics of the sugar inversion by Wilhelmy in 1850, the reaction rates of thousands of chemical transformations have been determined under the most varied conditions. The determination of the mechanisms of these reactions was the principal intent of these efforts, and the resolution of the overall reaction into the elementary processes was achieved in many instances. But such kinetic measurements cannot furnish any direct information about the end products and the yield of a reaction. Also, extensive studies of the appropriate conditions are necessary, especially for complex reactions. In addition, reproducible isothermal kinetic measurements require a careful temperature control and a corresponding technical effort. Then heat evolution complicates the kinetic investigation, even though it is a fundamental property of all reactions that deserve investigation in its own right. Both fundamental barriers for successful kinetic measurements—difficulties in preliminary work and errors due to heat—can be avoided by an unconventional mode of measurement and evaluation called the *nonisothermal reaction kinetics.*

The nonisothermal reaction analysis technique involves conducting a temperature programmed experiment. In order to observe the process from the onset, the temperature of the system should be increased continuously from a point where no reaction occurs. It has been shown that a linear increase in temperature with time is optimal in general, because then the simplest mathematical relationships between the signal and the reaction parameters prevail.

The analytical determination of the reaction-specific quantities $E$ and $A$ is based on experiments performed at varied temperature. The problem of obtaining conclusions to the processes involved from a continuously nonisothermal experimental curve can be solved by reconsidering kinetic laws in a different manner.

### 1.7.1 Constant Heating Rate Conditions

Both the mathematical treatment and the technical execution of a nonisothermal experiment are simplified if the temperature is increased by the experimenter from a starting temperature $T_s$ using a constant heating rate $\alpha$

$$T = \alpha(t_0 + t') = \alpha t \qquad (1.32)$$

with

$$t_0 = T_s/\alpha \qquad (1.33)$$

where

  $t$ = an imaginary time consumed if the heating were to begin at absolute zero

  $t'$ = time from the real start of the heating

Consider an elementary process:

  $A + (n - 1)A \rightarrow B$ or several products

  $n$ = reaction order (molecularity) $\qquad (1.34)$

When the reaction rate is followed, the following equation is obtained:

$$y = \frac{dx}{dt} = -\lambda \frac{d[A]}{dt} = \lambda k(T) \cdot [A]^n \qquad (1.35)$$

Here $y$ is called the reactivity, a function of the above type of reactivity function, and the corresponding curve $y = f(t)$ is the reactivity curve. If the Arrhenius law for the rate constant is valid, then $y$ is given by

$$y = \lambda A \exp(-E/RT) \cdot [A]^n \qquad (1.36)$$

or

$$y = \lambda A \exp(-\varepsilon/t) \cdot [A]^n \qquad (1.37)$$

with

$$\varepsilon = E/\alpha R = \text{relative activation energy} \qquad (1.38)$$

*First-Order Reaction*
For a first-order reaction, $n = 1$, and it follows that

$$[A] = [A]_0 \exp\left(-A \int_{t_0}^{t} e^{-\varepsilon/t} \, dt\right)$$

$[A]_0$ = initial concentration $\qquad (1.39)$

Using this value in the expression of $y$ and by making some approximations, it can be shown that

$$y = \lambda \cdot A[A]_0 e^{-\varepsilon/t - uk} = \lambda \cdot [A]_0 \cdot k(t) \cdot e^{-uk} \qquad (1.40)$$

where

$$u(t) = \frac{\int_0^t k \, dt}{k} \qquad (1.41)$$

The function $y/\lambda$ is called the first-order elementary function. Figure 1.3 gives the graphical depiction of first-order nonisothermal function.

The shape of the nonisothermal function is accounted for by the two variable factors: the first expresses the temperature influence from the Arrhenius law and increases to an asymptotical value $A$, the frequency factor; the second factor continually decreases to zero due to the consumption of the reactant.

*Second-Order Reaction*
In a bimolecular reaction of type

$$A + A \rightarrow B(+C) \qquad (1.42)$$

two molecules of the reactant are consumed in one elementary step. Thus the reactant concentration follows:

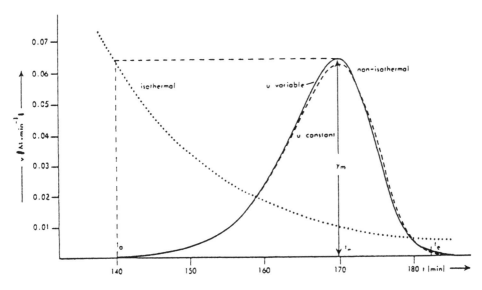

**FIGURE 1.3** Nonisothermal function for a first-order reaction.

$$[A] = \frac{[A]_0}{1 + 2[A]_0 uk} \tag{1.43a}$$

Hence, the second-order reactivity curve becomes

$$y = 2\lambda[A]_0^2 \frac{k}{(1 + a[A]_0 uk)^2} \tag{1.43b}$$

Several other equations have been developed to analyze the kinetic data obtained from constant heating rate experiments. Some of these are discussed in Chapter 4.

### 1.7.2 Changing Heating Rate Conditions

*Logarithmic Temperature Run*
For a logarithmic temperature run

$$-\frac{1}{T} = \ln t - a \tag{1.44}$$

The relationships for $y$ for first- and second-order reactions are, respectively, as follows:

$$y = \lambda A[A]_0 e^{-aE/R} t^{E/R} \exp\left[ -\frac{Ae^{-aE/R} t^{E/R+1}}{(E/R + 1)} \right]$$

$$y = \lambda \frac{A[A]_0^2 \exp\left(-\dfrac{aE}{R}\right) t^{E/R}}{[1 + A[A]_0 \exp(-aE/R) t^{E/R+1}/(E/R + 1)]^2} \tag{1.45}$$

However, these relationships are inconvenient because besides containing an exponential function, the activation energy is in the exponents of time.

*Exponential Temperature Run*

A program using exponential temperature increase would not be expedient because of further mathematical complications. Another situation is given in which instead of a temperature programmer, introduction of the sample into a reactor chamber at an elevated ambient temperature $T_u$ starts the reaction. If effects of reaction heat are eliminated, there is an exponential temperature increase:

$$T = T_u(1 - e^{-at}) \tag{1.46}$$

The first-order reactivity curve is transformed into

$$y = \frac{\lambda A}{a(T_u - T)} \exp\left(-\frac{E}{RT} - \frac{A}{a} \cdot I_1(T)\right) \tag{1.47}$$

where

$$I_1(T) = \int_0^T \frac{\exp\left(-\dfrac{E}{RT}\right)}{T_u - T} \, dT \tag{1.48}$$

For more details on nonisothermal reaction kinetics the reader should refer to Chapter 4 (or any other standard text book on non-isothermal reaction kinetics).

# 2

# OVERVIEW OF THERMAL ANALYSIS METHODS

## 2.1  INTRODUCTION

The name *thermal analysis* (TA) applies to all analytical techniques where the dependence of any physical property of a substance can be related to temperature measurements under controlled conditions. Analytical techniques such as proximate and ultimate analysis, thermometry, thermogravimetry (TG), differential thermogravimetry (DTG), differential thermal analysis (DTA), differential scanning calorimetry (DSC), and thermomechanical analysis (TMA) are all classified under thermal analysis (Wunderlich, 1990).

A given reaction can be followed either by keeping a record of *mass change* thermogravimetry and differential thermogravimetry, *heat exchange* differential thermal analysis and differential scanning calorimetry, or *dimensional change* thermomechanical analysis, as a function of time. The interrelationship of these various methods are illustrated in Figure 2.1.

Thermogravimetry involves the measurement of mass change during a reaction. The technique is applied to reactions where significant volatilization of sample mass occurs. Thermogravimetry was formerly known as thermogravimetric analysis, or TGA, but the International Confederation on Thermal Analysis recommends against using this term.

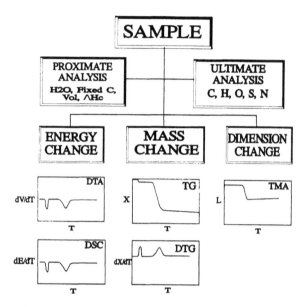

**FIGURE 2.1**   Relationship of various thermal techniques.

## 2.2   PROXIMATE ANALYSIS

The proximate analysis is a determination of moisture content, fixed carbon, volatile matter, and ash content of the sample. The carbon content determined through this method is not the actual carbon content but only the nonvolatile part of carbon content, as some of the carbon present in biomass also escapes along with the volatiles. This method, developed for coal, gives indication of char produced in a relatively simple and economic manner. It indicates the quality of coal and is used as an index for its characterization. It sometimes includes the heat of combustion of the sample [ASTM method 3173-87 (ASTM, 1989)].

## 2.3   ULTIMATE ANALYSIS

The *ultimate analysis* provides the composition of elemental carbon, hydrogen, oxygen, nitrogen, and sulfur for a combustible sample [ASTM D3177-48 (ASTM, 1989)].

A number of investigators have found that the elemental composition determined in the ultimate analysis is closely related to the heat of combustion (Tillman, 1978; IGT, 1978; Graboski, 1981). It has recently been found in a survey of these earlier works that the heat of combustion can be predicted from the ultimate analysis according to

$$HHV \ (kJ/g) = 0.3491C + 1.1783H + 0.1005S$$
$$- \ 0.1034O - 0.0151N - 0.0211Ash \qquad (2.1)$$

where HHV is the high heating value of the sample, and C, H, S, O, N, and Ash are the weight % of carbon, hydrogen, sulfur, oxygen, nitrogen, and ash, respectively (Channiwala, 1992). The average absolute error of prediction is 1.45%, better than that claimed for other correlations, and the bias error for all compounds was found to be negligible, which can be approximated to 0%. The accuracy of the ASTM bomb calorimetry method has a reproducibility limit of 240 J/g, while this correlation for the entire range of fuels offers predictions within 337 J/g, quite comparable with the measurement uncertainties, particularly considering the widely varying nature of data and the sources of their collection. The proximate, ultimate analysis, and heat of combustion data on biomass and some fuel materials samples taken from this work are listed in the appendix.

## 2.4 THERMOGRAVIMETRY AND DIFFERENTIAL THERMOGRAVIMETRY

*Isothermal thermogravimetry* deals with the change in sample mass at constant temperature as a function of time and is used to measure rate constants. In this process, a known amount of sample is placed in a furnace in a controlled gaseous atmosphere at the desired temperature. The weight loss of the sample is recorded as a function of time. The graph of weight loss versus time is referred to as a thermogram. It is used to analyze the thermal stability of the sample, the rate of decomposition, the effect of various gases on the decomposition of a sample, and possible adsorption and desorption reactions, and to determine the kinetics of the reactions occurring.

*Dynamic thermogravity* is the most common form of thermogravimetric analysis. In this technique, a continuous recording of weight change of the sample in a flowing or static gas atmosphere is made as a function of time or temperature at a fixed heating rate and plotted against temperature. Modern instruments can vary the heating rate over a wide range and hold at various temperatures for specified times, for instance to dry the sample prior to making a test run.

Since weight and temperature data are continuously recorded during the course of a TG measurement, it is a simple matter of computation to show the first derivative of weight versus time or temperature, emphasizing the zones of reaction. This method of TG data reporting is known as differential thermogravimetry (DTG).

TG and DTG have several advantages over isothermal thermo-gravimetry (Cardwell and Lunar, 1976). One of the major advantages is that, by conducting one experiment, it is possible to characterize the behavior of a specimen over a very wide temperature range, e.g., 25–1000°C, thus providing an overview of thermal behavior at all temperatures of interest. It is also possible to determine kinetic parameters of the various reactions over a much wider temperature range than can be studied in isothermal TG by varying the heating rate. Other advantages are elimination of error from sample to sample as only one run is used to determine the kinetic parameters and the reduction of time in performing the kinetic studies.

## 2.4.1   Applications of Thermogravimetric Analysis

Modern thermogravimetry started in the late 1950s with the availability of high-quality thermobalances. Now TG is almost universally applied in the fields of metallurgy, ceramics, inorganic and organic chemistry, polymers, biochemistry, geochemistry, forensics, among others. Some of the applications of thermogravimetry are thermal decomposition of organic, inorganic, and polymeric substances in an inert atmosphere to help in determining the pyrolysis kinetics of decomposition or to determine the thermal stability of the sample; reaction chemistry of solid samples where one of the products is in a gaseous state; distillation and evaporation of liquids; proximate analysis; vapor pressure determinations; and evaporation and sublimation.

Automatic thermogravimetric analysis developed by Duval (1951) determined metallic ion or mixture of ions by the use of thermogravimetry. In this method, the crucible is loaded with a known amount of precipitate of a mixture. It is then heated in a controlled manner and the mass loss curve is recorded. Once a horizontal plateau is reached, the mass of the leftover precipitate is obtained. This leftover precipitate is further heated until another plateau is obtained at higher temperature, where once again the mass of the remaining precipitate is obtained. These mass levels help indicate a definite stoichiometry of the precipitate (Duval, 1951).

Griffith (1957) has applied thermogravimetry to the determination of the moisture content in the hydrated and anhydrous salts in different phases. The method is based on the fact that under controlled heating rates, a selective decomposition of the phase with the highest dissociation temperature takes place. When one phase is decomposed, the phase with the next highest decomposition temperature decomposes, and so on.

Hoffman (1959) applied thermogravimetry to the analysis of clay and soils. He was able to determine water content, organic content, and the inorganic carbonates from pure clays and mixture of clays. The results obtained from TG were in good agreement with those found by X-ray diffraction and wet chemical analysis.

TG analysis has also been applied to polymers. Application of TG in this field includes study on the relative thermal stability, the effect of additives on the thermal stability, decomposition kinetics, quantitative analysis of various copolymers, oxidation stability, among others. TG has been used to provide information regarding the presence of additives such as plasticizers in the polymer. A thermogram of pure poly vinyl butyryl was compared with one containing plasticizer by Wendlandt and Barbson in 1959 (Wendlandt, 1974). The difference between the two thermograms helped in the quantitative determination of plasticizer content.

Since the 1960s, the use of thermogravimetry extended to the determination of kinetic parameters for various gas-solid reactions. Several methods have since been developed to analyze TG data for quantitative estimate of kinetic parameters. Some of these methods have been discussed in Chapter 4. Recent application of TG has been for the deconvolution of biomass components such as hemicellulose, cellulose, and lignin (Varhegyi et al., 1993). This technique is still at the developmental stage.

## 2.5   DIFFERENTIAL THERMAL ANALYSIS AND DIFFERENTIAL SCANNING CALORIMETRY

Some reactions are endothermic or exothermic but may not involve a weight change and hence cannot be studied by thermogravimetric analysis. The technique of differential thermal analysis (DTA) or differential scanning calorimetry (DSC) can be applied in such cases for determining reaction kinetics by detecting endothermicity or exothermicity of reactions.

A technique similar to DTA was first used by Le Chatelier in 1887 for the identification of clay materials and pyrometry. He introduced heat curves for the identification of different clays. This analysis eventually gave rise to differential thermal analysis technique. In his study for the identification of different clay materials, Le Chatelier used the rate of deceleration in temperature rise at the point of dehydration, e.g., he found that crystalline kaolin from Red Mountain, Colorado, has a deceleration point at 770°C and 1000°C. This characteristic deceleration in temperature was also found in kaolin from places like France and China. It was observed by Le Chatelier that the adsorption and evolution of heat for a given sample as a function of temperatures is a unique property of that material (Le Chatelier, 1887).

*Differential thermal analysis* is a technique of recording the difference in temperature between a substance and a thermally inert reference material as the two specimens are subjected to identical temperature treatment in an environment heated or cooled at a controlled rate. In this technique, the temperature of a sample is compared to the temperature of the thermally inert material as both are heated at a linear heating rate. It is a differential method in the sense that the temperature of the sample $T_S$ is compared to the temperature of the reference material $T_r$ and the difference between the two is used to determine the heat of reaction. The reference material is usually calcined alumina or a similar inert material. It is known that there are no phase transformations for calcined alumina up to the temperature of 1100°C. This ensures that the energy changes that are recorded will be due to the transformations taking place in the sample under consideration. DTA does not spell out what net reaction is taking place when an exothermic or endothermic effect is recorded. It simply records that there is a change in energy content taking place and the rate of that change.

The reactions that can be reflected by the DTA curve are phase transitions; solid state reactions; decompositions; surface reactions and second-order transitions; and a change in entropy without a change in enthalpy.

Since DTA is based on the measurement of temperature changes in relation to a reference material, the choice of the thermocouple and thermocouple placement is an important aspect of the design of DTA instruments (Murphy, 1958).

The amount of reference material used is such that the total heat capacity or mass of the reference material and the sample are about the same, in order to avoid any thermal lags between the two thermocouples during the heating of the furnace. Whenever there is a transition taking place in the sample, the temperature of the sample increases less or more rapidly than that of the reference, depending on the endothermic or exothermic process. This change in heating rate is recorded in terms of temperature difference between the reference and the sample thermocouple. The position of the peaks is related to the chemical change of the substance and the area of the peak is proportional to the energy involved in the reaction occurring. In cases where the entire or the major portion of the sample vaporizes during transition, there can be large deviation between the reference and the sample due to the change in heat capacity of the sample. In such cases it is recommended that the sample be doped in the reference material, e.g., alumina in small quantity so that the heat capacity of both the reference and the sample pans are about the same due to the presence of alumina

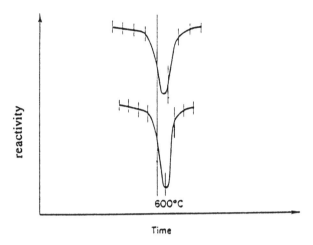

**FIGURE 2.2** A typical endothermic peak obtained during DTA experimentation.

in both the pans during the entire range of temperature of interest. This procedure limits the temperature deviation caused due to heating of the sample to a negligible value.

A typical DTA curve is shown in Figure 2.2, showing an endothermic peak. The point where a tangent to the side of the peak and to the base line meet is the commencement of the peak. The base lines ideally should be aligned, but in practice there is a drift with increasing temperature, due to changes in thermal conductivity or mass.

One thing which is evident from examining the endotherm and exotherm is that the heating rate within the sample during the occurrence of a reaction is not the same as that when no reaction is occurring nor is it uniform. This is because the sample temperature lags from that of the reference material, which causes the deviation in temperature between the two pans.

*Differential scanning calorimetry* is another technique for measuring heats involved in phase changes. In DSC, a pilot heater is placed around the sample holder and heat is added or subtracted in small quantities to keep the temperature of the sample and the reference the same. The amount of heat required is monitored and presented graphically. Typical accuracies of heat measurements range from ±10% to a few tenths of a percent depending on the heating rate employed.

## 2.5.1 Applications of Differential Thermal Analysis

The merit of the DTA curve is that all energy changes occurring in the sample during heating are clearly observable (provided the sensitivity is high enough) and that the peak areas reflect the energy involved. It can be said with certainty that no two materials give exactly the same complete DTA curve since even slight differences in chemical composition or crystal structure are reflected in differences in either the temperature at which the reaction occurs or the heat of reaction. Because of this property, each substance gives a unique DTA curve for a particular sample. By the selection of a limited number of substances having DTA characteristics very similar to that of unknown samples, it is possible to deconvolute the properties of unknown samples. When this technique is applied to mixtures, sometimes it is possible to interpret the entire DTA as the sum of pure components. Therefore, it is advisable to have some idea about the origin of the unknown sample. This limitation of DTA suggests that in itself alone it is not a good identification technique and that the results should be interpreted in conjunction with other techniques, such as X-ray diffraction or spectroscopic analysis.

Roberts-Austen in 1899 suggested the use of reference inert material to measure the differential temperature. This method was first tested on different metals. Fenner first applied this technique for the identification of clay and silica minerals in the United States, and it soon became the standard technique for the identification of different clays. Hollings and Cobb (1923) introduced the control of gaseous atmosphere around the sample.

Norton (1939) conducted studies on the identification of different clays by using DTA. He found a relationship in the area of the endotherm versus the weight percent of a mineral in the mixture indicating proportionality between the weight of the mineral and the heat effect.

Wittels (1951) used DTA as a microcalorimeter. He found that the area of the endotherm is linearly related to the heat of reaction for given mass of the sample. Around the same time, Berger and Whitehead (1951) used DTA to investigate the coalification process. In their study, they find that lignin gives an exothermic peak between 400 and 500°C, while cellulose yields an endothermic peak at 400°C. Measurements on a variety of samples show that the lignin peak is identifiable in bituminous coals and lignites. The peak is absent for anthracites, indicating the complete conversion of the lignin type structures to thermally more stable forms at the end of coalification. Indications for small amounts of cellulosic structures were found in peat and lignite but they were absent in bituminous coals. Based on this lignin theory, a coal formation was reemphasized.

An extensive investigation of the relationship between coal rank and the characteristics of DTA curve was made by Glass

(1955). The DTA curves are classified into five distinct types: (1) the meta anthracite type, characterized by single endothermic peak at 725–735°C; (2) the anthracite type showing a single volatile loss peak at 630–680°C; (3) the low volatile bituminous type endotherm at 500–620°C; (4) the high volatile bituminous type, where primary volatilization endothermic effect is divided into two sharp exothermic reactions; and (5) the sub-bituminous type having a large primary volatilization peak at 450°C.

Garn and Flaschen (1957) report the use of DTA for the analysis of impurities in mineral matter. Schwenker and Beck in 1960 presented a study on the detection of transition temperatures for textile fibers (Mackenzie, 1970).

Jen Chiu (1962) has shown that DTA can be used to identify the organic compounds. Sanderman and Augustine (1963) applied DTA to determine the thermal stability for different woody components. The order they rank is lignin > cellulose > hemicellulose. Hemicellulose decomposed exothermically at about 200°C in nitrogen. However, some polysaccharides would decompose closer to cellulose decomposition temperature. The thermal stability of polysaccharides and monosaccharides is reduced by the presence of carboxyl groups. Cellulose decomposes with an endotherm commencing around 290°C. Sometimes an exotherm is seen at around 350°C. The lignin has a flat exotherm peak commencing at 300°C and reaching a maximum at 425°C.

DTA can easily detect the stability ranges for a sample. One of the major advantages is that the results can be obtained with much rapidity.

DTA has proven to be of great value for the analysis of metastable and unstable systems. A metastable system has a higher free energy than the corresponding equilibrium system, but does not change noticeably with time. An unstable system in contrast is in the process of changing towards the equilibrium and can only be analyzed as a transient state. Metastable states usually become unstable during DTA, which operates without temperature gradients in the sample and reference. If a sizable temperature gradient exists within the sample then the sensitivity may be higher than with smaller gradients because one can work with larger masses, but for quantitative heat measurements it is desirable to have as little temperature gradient in the sample as possible. Samples such as some oxides and organic materials have lower thermal conductivity and therefore may have internal temperature gradients. Detailed history of DTA development and its applications is given by Mackenzie (1970).

Since the early 1970s, the total number of research papers involving DTA exceeded 1000 per year; it is therefore hard to keep account of these works. However, no major change took place in the development of DTA techniques as such; the only change that did take place was in terms of application of DTA in different research areas such as polymer characterization, deconvolution of solid waste streams, catalyst development, and fuel sciences in general. Currently, most of the new instruments are being made to combine TG-DTA-MS to have better understanding of the overall reaction process.

## 2.6  ICTA RECOMMENDATIONS

Due to the ever-increasing use of TG-DTA techniques, in 1965 the first International Conference on Thermal Analysis (ICTA) established a committee on standardization charged with the task of studying how and where standardization might further the value of thermal analysis. One area of concern was for the uniform reporting of data, in view of the profound lack of essential experimental information occurring in much of the thermal analysis literature. Because thermal analysis usually involves dynamic techniques, it is essential that all pertinent experimental details accompany the actual experimental records to allow their critical assessment. This was emphasized by Newkirk and Simmons (Wunderlich, 1990), who offered some suggestions for the information required with the curves obtained by thermogravimetry.

The actual format for communicating these details will, of course, depend on a combination of the authors' preference, the purpose for which the experiments are reported, and the policy of the particular publishing medium. To accompany each TG or DTA record, the following information should be prepared:

1. Identification of all substances (sample, reference, diluent) by a definitive name, an empirical formula, or an equivalent compositional data list.
2. A statement of the source of all substances, details of their histories, pretreatments, and chemical purities, so far as these are known.
3. Measurement of average rate of linear temperature change over the temperature range involving the phenomena of interest.
4. Identification of the sample atmosphere by pressure, composition, and purity whether the atmosphere is static self-generated or dynamic through or over the sample. Where applicable, the ambient atmospheric pressure and humid-

ity should be specified. If the pressure is other than atmospheric, full details of the control methods should be given.

5. A statement of the dimensions, geometry, and materials of the sample holder, the method of loading the sample holder, and the method of loading the sample, where applicable.

6. Identification of the abscissa scale in terms of time or of temperature at a specified location. Time or temperature should be plotted to increase from left to right.

7. A statement of the methods used to identify intermediates or final products.

8. Faithful reproduction of the original records.

9. Wherever possible, each thermal effect should be identified and supplementary supporting evidence stated.

In the reporting of TG data, the following additional details are also necessary:

10. Identification of the thermobalance, including the location of the temperature measuring thermocouple.

11. A statement of the sample weight and weight scale for the ordinate. Weight loss should be plotted as a downward trend, and deviations from this practice should be clearly marked. Additional scales (e.g., fractional decomposition and molecular composition) may be used for the ordinate where desired.

12. If derivative thermogravimetry is employed, the method of obtaining the derivative should be indicated and the units of the ordinate specified.

In reporting DTA traces, these specific details should be presented:

13. Sample weight and dilution of the sample.

14. Identification of the apparatus, including the geometry and materials of the thermocouples and the locations of the differential and temperature measuring thermocouples.

15. The ordinate scale should indicate deflection per degree centigrade at a specified temperature. Preferred plotting will indicate upward deflection as a positive temperature differential, and downward deflection as a negative temperature differential, with respect to the reference. Deviations from this practice should be clearly marked. We have tried to follow these recommendations in this atlas.

## 2.7 MANUFACTURERS

Commercial DTA first became available in 1952 from the Robert L. Stone Company. Stone's was the first quantitative application of DTA to a chemical system in the study of the polymorphism of $Na_2SO_4$ by Kracek in 1929. In this study, temperatures for five different phase transformations were identified. Stone determined the heat of dissociation of magnesite to be 10.1 kcal/mole. The thermogram showed increasing decomposition temperature with increasing $CO_2$ pressure. When $\log p$ is plotted against $1/T$, the slope of the plot is $\Delta H/R$ and can be used to determine the heat of reaction from the Clausis-Clapyron equation, i.e.

$$d(\ln P)/d(1/T) = -Z(\Delta H/R) \qquad (2.2)$$

where

$Z$ = compressibility factor
$P$ = pressure
$\Delta H$ = heat of reaction
$T$ = temperature
$R$ = gas constant

There are several commercial manufacturers of TG equipments. Prominent among those are Perkin-Elmer, Du Pont, Stanton Redcroft, Seiko Instruments, Cahn, Sinku Rico, and Mettler. The differences in the design of various equipment in terms of accuracy for parameter measurements is mainly in the placement of the thermocouple for the measurement of sample temperature or the mechanism of gas solid contact, i.e., whether the gas flows over the sample, flows through the sample, or diffuses into the sample. Difference can also be seen in terms of sample holder to sample size ratio. This aspects relates to the temperature deviation within the sample at given heating rates.

The operating limits on these models also varies. Instead of discussing each model, we present the extreme limits available. The maximum sample size that can be subjected to thermal decomposition is 10 g in the Cahn TG-131 model. (This thermobalance can hold a sample size up to 100 g, but the dynamic range is only 10 g.) All the other models restrict maximum weight up to 200 mg. The maximum heating rate of 100°C/min is available in almost all of the commercial models. However, the slowest heating rate, of 0.1°C/min, is available in the Seiko SSC 5200 model. A pressure variation from 5 torr to ambient is available in most of the commercial designs. Only the Cahn TG-151 has the capability to go up to 1000 psig. The determination of temperature calibration by using Curie point method is easiest in Perkin-Elmer model TG-7.

The data presented in this book were collected on a Seiko Instruments TG/DTA SSC 5200 model. One of the major advantages of this unit is that it can collect data on the same sample simultaneously for TG and DTA. This feature is not present in most other commercially available units. We will discuss the details of this instrument and the procedure for data collection in Chapter 5. Some other models can record data for TG and DTA simultaneously (Cahn TG-131/DTA-131) but they require a different sample for TG and DTA recordings.

# 3

# DERIVING KINETIC DATA FROM THERMAL ANALYSIS

## 3.1   INTRODUCTION

The determination of kinetics of decomposition for biomass and other fuels from thermal analysis data have long been a source of fascination to those working in this field. It is projected that an accurate determination of the kinetic parameters from TG data for a particular biomass material would be useful in designing process plants. However, to the best of our knowledge, the existing kinetic parameters determined from this technique have never been applied because there was such a scatter in kinetic data that no one knew which set of data appropriately explains the reaction process.

It is a relatively simple matter to fit TG curves with an equation of the following form (Milne, 1981):

$dV/dt = kV^n$

where

$k = A \exp(-E/RT)$
$V$ = fraction of total volatiles remaining at temperature $T$
$n$ = order of reaction
$E$ is sometimes interpreted as an activation energy
$A$ = pre-exponential factor
$R$ = gas constant

If the sample is heated at a constant rate, $a = dT/dt$, then this equation becomes

$$dV/dT = kV^n/a$$

However, when it comes to comparing data from one research work to the other, a wide variation is found in the kinetic parameters for even biomass components like cellulose, which have relatively fixed structure. A number of workers have used cellulose to determine the decomposition kinetics and decomposition mechanisms (Friedman, 1965; Chatterjee, 1965; Broido, 1969; Shafizadeh, 1979; Diebold, 1985; Agrawal, 1986). In particular, Antal has written two extensive reviews on the decomposition kinetics of cellulose and lignocellulosics based on various proposed reaction schemes involving up to five steps (Antal, 1982, 1985). More recently, Varhegyi, Jakab, and Antal have published another review that suggests that a single step is sufficient for describing the kinetics of the cellulose decomposition (Varhegyi et al., 1994).

Cellulose comes close to being an identifiable "biomass compound" with potentially reproducible kinetics. Yet the kinetic exponential factor, $E$, has been reported to vary from 50 to 250 kJ/mole. As shown in Figure 3.1, an excellent correlation exists between the frequency factor and the activation energy. This has been called the "compensation effect" (Chornet, 1979; Reed, 1985). Much of this variation is illusory; TG data plotted from any of the pairs of $A$ and $E$ in Figure 3.1 shows cellulose breaking down at temperatures between 280°C and 400°C with varying slopes. This variation

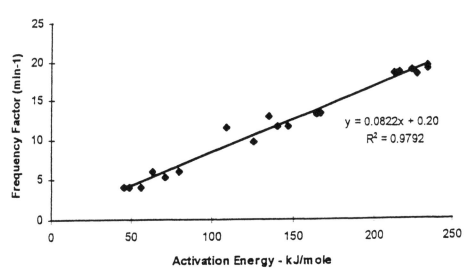

**FIGURE 3.1** Compensation effect for cellulose kinetics.

in the kinetic factors can be attributed to variation in sample size and heating rate (discussed in Chapter 13), variations in the methods used for mathematical analysis (this chapter), and the instrumental errors coupled with sample impurities (Chapter 5).

Recently there has been some progress in determining the sources of the discrepancies and deriving reliable data (Antal et al., 1994; Fritsky et al., 1994; Gaur and Reed, 1994; Varhegyi et al., 1993; Abatzoglou et al., 1992; Belkacemi et al., 1991). We hope that the data in this book along with these recent developments in kinetic interpretations can provide useful data for future analysis and lead to a better understanding of the kinetics of thermal decomposition for biomass and other solid fuels, which can then be applied for the practical design of process plants.

For those wishing to do kinetic analysis, we have tried to put together some of the different methods available in thermal analysis literature (TG and DTA) for the interpretation of data to determine the kinetic parameters. We are not, however, attempting to rank these methods because we believe that the readers of this book can make the judgment as to which method will satisfy their needs in the best way.

## 3.2 DETERMINATION OF KINETIC PARAMETERS USING THERMOGRAVIMETRIC DATA

Several mathematical methods have been developed to determine the kinetic parameters from the data obtained during TG experimentation. These methods have been developed over the years since the early 1960s and are still evolving. In this section, we present a brief review of the methods that are either in common use among the researchers or have provided a major change in the way TG data should be analyzed.

### 3.2.1 Coats and Redfern Method

For a reaction $A_{(s)} \rightarrow B_{(s)} + C_{(g)}$, the rate of conversion of $A$ can be given as

$$dX/dt = k(1 - X)^n \tag{3.1}$$

where $X$ is the fractional conversion and $k$ is the rate constant, expressed as

$$k = A \exp(-E/RT) \tag{3.2}$$

$A$ and $E$ are Arrhenius parameters known as the frequency factor and the activation energy, respectively, while $n$ is the reaction order

(Coats and Redfern, 1964). Since not much theoretical justification can be provided for frequency factor and reaction order, in the case of solid decomposition, Gadalla (1985) recommends the use of terms *pre-exponential factor* and *exponent term*, respectively, to avoid any misconception. In this book we have tried to adhere to this terminology.

For a linear heating rate, $a = dT/dt$, the substitution for $k$ in equation (3.1) can be made as follows:

$$dX/dT = A/a \exp(-E/RT)(1 - X)^n \tag{3.3}$$

or

$$dX/(1 - X)^n = A/a \exp(-E/RT)\, dT \tag{3.4}$$

In order to integrate equation (3.4) within limits of $X = 0$ to $X$ and $T = 0$ to $T$, at first substitute $u = E/RT$. The approximate solution for the right-hand side of this integral is then given as

$$\int e^{(-u)}u^{(-b)}\, du = u^{(1-b)}e^{-u} \sum (-1)^n (b)^n/(u^{(n+1)}) \tag{3.5}$$

Substituting this solution into equation (3.4) for $n$ not equal to 1 gives

$$1 - (1 - X)^{(1-n)}/(1 - n) = ART^2/aE[1 - 2RT/E]e^{-E/RT} \tag{3.6}$$

and for $n = 1$, equation (3.4) becomes

$$-\log[(1 - X)/T^2] = AR/aE[1 - 2RT/E]e^{-E/RT} \tag{3.7}$$

To determine the kinetic parameters for any reaction by using this method, one has to make a guess at the reaction order. Subsequently, by taking the log of both sides, equations (3.6 and 3.7) transform into the linear function $y = mx + c$. The reaction order for which the linearity of the function is best defined can then be used to determine the kinetic parameters by using the graphical technique.

This method of Coats and Redfern (1964) has been used successfully by many researchers. However, one of the drawbacks of this method is that the reaction order has to be known beforehand or one has to do extensive linear fit on the experimental data for various reaction orders. Another limitation is that for cases where there is stepwise conversion, such as for calcium oxalate or polyvinyl chloride, the treatment of the data has to be done for separate steps individually.

## 3.2.2 Gyulai and Greenhow Method

This method considers decomposition of solids at two heating rates for the prediction of the kinetic parameters. By doing so, the authors have eliminated the requirement of assuming the rate function or the reaction order as is the case with the Coats and Redfern method.

The basic rate equation used by Gyulai and Greenhow (1974) is similar to equation (3.1) except that they do not define the form of the solid conversion function

$$\int dX/dT = \int A/a \, \exp(-E/RT) f(X) \tag{3.8}$$

In this case the solution is even more complicated because if we separate the variables

$$\int dX/f(X) = \int A/a \, \exp(-E/RT) \, dT \tag{3.9}$$

then the unknown term is present on both sides of the equation; the right-hand side of the expression does not have an exact integral solution, and on the left the form of the function for the conversion of the solid is not known.

By substituting $F(X) = \int [d(X)/f(X)]$ in equation (3.9), we obtain

$$F(X) = A/a \int [\exp(-E/RT) \, dT] \tag{3.10}$$

If $i = \int [\exp(-E/RT) \, dT]$, then

$$F(X) = (A/a)(i) \tag{3.11}$$

For two points corresponding to two identical conversion extents $X$, on two TG curves obtained at two different heating rates $a_1$ and $a_2$ for the same sample and initial weight, we can write

$$i_{11} = \int [\exp(-E/RT) \, dT] \tag{3.12}$$

$$i_{12} = \int [\exp(-E/RT) \, dT] \tag{3.13}$$

$$F(X)_{11} = (A/a_1)(i) \tag{3.14}$$

$$F(X)_{12} = (A/a_2)(i) \tag{3.15}$$

where the first figure on the subscript relates to the conversion extent and the second value relates to one of the two TG curves.

Since the points have been obtained for the same conversion level and the sample is the same, they would in all probability exhibit the same conversion function, and hence

$$F(X)_{11} = F(X)_{12} \tag{3.16}$$

It follows that

$$F(X)_{11}/F(X)_{12} = (a_2/a_{11})/(i_{11}/I_{12}) = 1 \tag{3.17}$$

or

$$a_1/a_2 = I_{11}/i_{12} = \cdots = i_{nj} = i_{nk} \tag{3.18}$$

For the determination of the activation energy, two points of the same conversion are chosen on two TG curves obtained at two different heating rates and the corresponding temperatures $T_{11}$ and $T_{12}$ are noted. The value of $\log i$ corresponding to these temperatures at various activation energies can be obtained from the tabulated values given by authors, or from any of the approximate solutions such as the one given by Coats and Redfern. The plot of $\log(i_1/i_2)$ against $E$ would help in finding that value of $E$ corresponding to which $\log(a_1/a_2) = \log(i_1/i_2)$.

Now for the two points on same TG curve, we can write

$$F(X_1) = (A/a)i(E_1, T_1) \tag{3.19a}$$

$$F(X_2) = (A/a)i(E_2, T_2) \tag{3.19b}$$

Then, depending on the type of conversion function, say $F(X) = \int dX/(1 - X)^n$, we want to explore, the integral solution of the function can be obtained, for example in this case for $n$ not equal to 1

$$F(X) = \left[\frac{1}{(n - 1)}\right]\left[\frac{1}{(1 - X)^{n-1}} - 1\right] \tag{3.20}$$

and for $n = 1$

$$F(X) = -\ln(1 - X) \tag{3.21}$$

Now we choose one reference conversion, say $X_2$, and calculate the ratio $\log[F(X_1)/F(X_2)]$ for a range of values from the conversion $X$ data. This ratio is then plotted against $n$ using equation (3.20). The order of the reaction for a conversion $X_1$ is obtained from the graph as the value of $n$ corresponding to $X = X_1$ and $\log[F(X_1)/F(X_2)]$ $= \log[i(E_1, T_1)/i(E_2, T_2)]$. After these determinations, the pre-exponential factor is calculated with the help of equation (3.11).

### 3.2.3 Doyle's Method

Doyle's method (Doyle, 1961) deals with the decomposition in each step as being independent. The apparent rate of volatilization is found by

$$\frac{dV}{dt} = -a \left( \frac{dW}{dT} \right) \tag{3.22}$$

where

$W$ = initial weight
$a$ = heating rate
$V$ = volatile portion of the solid

However, if there are different volatilization steps, then for a particular step the appropriate residual mass fraction, $h$, is calculated on the total fraction volatilized during the step, rather than on the total initial mass:

$$h = \frac{(W - G)}{H} \tag{3.23}$$

where

$H$ = total mass fraction volatilized during the step
$G$ = weight fraction remaining after the step has been completed

Hence, we can now write

$$\frac{dV}{dt} = H \left( \frac{dh}{dt} \right) \tag{3.24}$$

We can define $dh/dt$ with a kinetic expression

$$-\frac{dh}{dt} = kf(h) \tag{3.25}$$

where

$k = A \exp(-E/RT)$
$f(h)$ = function for solid conversion

The conversion of the solid for a given step has been written by Doyle (1961) in the same manner as by others

$$\int \frac{dh}{f(h)} = \frac{A}{a} \int \left[ \exp \left( -\frac{E}{RT} \right) dT \right] \tag{3.26}$$

The solution of the right-hand side of the equation is done by substituting $u = E/RT$ and then obtaining the approximation $p(x)$ for the above integral:

$$\int \frac{dh}{f(h)} = g(x) = \left( \frac{A}{a} \right) \left( \frac{E}{R} \right) p(x) \tag{3.27}$$

where

$$\log p(x) = -2.315 - 0.4567 \, E/RT \tag{3.28}$$

Doyle points out that the value of $A/a$ can be evaluated on the basis of a single thermogram slope $dV/dT$ and the corresponding absolute temperature $T$:

$$\frac{A}{a} = \left[ \frac{\exp\left(\dfrac{-E}{RT}\right)}{Hf(h)} \right] \left(\frac{dV}{dT}\right) \qquad (3.29)$$

The value of $E$ is evaluated by first determining $X_a$, the value of $X$ at the corresponding value of $T_a$:

$$E = \left[ \frac{RHf(h_a)g(h_a)\exp(-X_a)}{p(x_a)} \right] \left(\frac{dT}{dV}\right)_a \qquad (3.30)$$

Doyle's method was one of the earliest methods developed for the determination of the kinetic parameters using nonisothermal TG data.

## 3.2.4 Zsako Method

Zsako (1970, 1973) has considered a more general form of rate expression than the methods discussed up to now in this chapter:

$$\frac{dX}{dt} = kX^a(1 - X)^b \qquad (3.31)$$

where $a$ and $b$ are empirical constants.

In this method, Zsako has tried to provide an improvement in terms of applying Doyle's method for reactions that have more complex conversion functions, $f(x)$. Taking the log of equation (3.27) gives

$$\log\left[ \left(\frac{A}{a}\right) \left(\frac{E}{R}\right) \right] = B = \log g(X) - \log p(X) \qquad (3.32)$$

Now once again taking the log of the left side of equation (3.32)

$$\log A = B + \log(Ra) - \log E \qquad (3.33)$$

Zsako has determined the analytical forms of conversion function $(1 - X)^b$ and term $B$ for reaction orders (b) e.g., 0, 1/3, 1/2, 2/3, 1, and 2. In this method, the function with the most constancy over the entire thermogram can be considered the conversion function of the solid. The value of $p(x)$ is given by Doyle as a function of activation energy.

### 3.2.5   Satava and Skvara Method

The method of Satava and Skvara (1969) is basically a more general form of the method proposed by Zsako. The rate expression used to describe the solid decomposition by Zsako is given as equation (3.31).

At a constant heating rate, the decomposition of a solid can be written as in equation (3.9). The integral of the left side of this equation can be performed once the decomposition function for the solid is known. In this method, since generalized forms of the function have been considered, no single solution is available. The right-hand side has been integrated by Doyle and many others. The authors of this method preferred to use the solution provided by Doyle (equation 3.27). Taking logarithms of both sides of equation (3.27) gives equation (3.32). It can be seen that quantity $B$ is independent of temperature.

Satava and Skvara determined the values of $\log g(X)$ for various functions. The major advantage of this method lies with the fact that the authors have considered and provided solutions to the different forms of conversion functions.

Now, since both the terms on the right-hand side of equation (3.32) are known, the values for $B$ can be determined. Satava and Skvara propose a graphical technique for the determination of value for $B$. The values of $X$ at intervals of 0.05 and the corresponding temperature $T_X$ are first read from the TG curve. The $\log g(x)$ values for the various rate processes are plotted against the corresponding $T_X$ values on transparent paper. On the same scale, a plot of $-\log p(x)$ versus $T$ is also drawn. The plot of $\log g(x)$ is placed on top of the $\log p(x)$ diagram so that the temperature scales coincide, and is then shifted along the ordinate until one of the $\log g(x)$ curves fits one of the $\log p(x)$ curves. From this $\log p(x)$ curve, the activation energy $E$ is determined. The $\log g(x)$ curve that fits the $\log p(x)$ curve is the most probable kinetic function describing the thermal decomposition of that particular solid. When the two plots are placed so that the values on the axes coincide, the distance between the $\log g(x)$ and $\log p(x)$ curves is the value for $B$, which is used to quantify pre-exponential factor $A$ by using equation (3.31).

It is, however, advised that in order to be fully confident of the conversion mechanism determined by this method one should also check it with some other method apart from TG analysis to rule out the possibility of mathematical coincidence of curve fitting.

## 3.2.6 Freeman and Carroll Method

The Freeman and Carroll method helps in determining the kinetics of a reaction over the entire range of temperature (Freeman and Carroll, 1958). The kinetic parameters are determined by using equation (3.34):

$$\frac{\Delta \log a \left(\frac{dC}{dt}\right)}{\Delta \log(1 - C)} = n - \left(\frac{E}{2.3R}\right) \frac{\Delta \left(\frac{1}{T}\right)}{\Delta \log(1 - C)} \tag{3.34}$$

This method is also called a difference differential method. A plot of $\Delta \log(a \; dC/dt)/\Delta \log(1 - C)$ versus $\Delta(1/T)/(\Delta \log(1 - C))$ is used to determine the value for activation energy, and subsequently equation (3.34) gives the reaction order. It has been found that by the use of this equation the results are in good agreement at low conversion levels.

## 3.2.7 Ingraham and Marier Method

Ingraham and Marier developed a method for reactions that follow linear kinetics for the decomposition, such as calcium carbonate. The rate expression for such a reaction can be written as

$$\frac{dw}{dt} = k \tag{3.35}$$

where $dw$ = mass loss per unit area in time $dt$. If the temperature of the sample is increased at a linear heating rate, the temperature at any given time can be given as

$$T = b + at \tag{3.36}$$

where

$b$ = initial temperature
$a$ = heating rate

These authors developed equation (3.37) for the determination of activation energy

$$\log \left(\frac{dw}{dT}\right) = \log T - \log a + \log C - \left(\frac{E}{2.3R}\right) \tag{3.37}$$

The activation energy is calculated from the slope of a plot of $[\log(dw/dT) - \log T + \log a]$ versus $1/T$. The $\log(a)$ value permits the correction of TG curves obtained at different heating rates. We will discuss the correction of TG curves at different heating rates in greater detail later in this chapter.

### 3.2.8   Vachuska and Voboril Method

This is a differential method for the determination of kinetic parameters using thermogravimetric data:

$$\frac{dX}{dt} = k(1 - X)^n \exp\left(-\frac{E}{RT}\right) \tag{3.38}$$

or

$$\ln\left(\frac{dX}{dt}\right) = \ln k + n \ln(1 - X) - \left(\frac{E}{RT}\right) \tag{3.39}$$

Since $X$ and $T$ are functions of time in the case of TG data, we can differentiate the above equation with respect to time

$$\frac{\left(\dfrac{d^2X}{dt^2}\right)}{\left(\dfrac{dX}{dt}\right)} = -\left(\frac{n}{1 - X}\right)\left(\frac{dX}{dt}\right) + \left(\frac{E}{RT^2}\right)\left(\frac{dT}{dt}\right) \tag{3.40}$$

Rearrangement of equation (3.40) gives

$$\frac{\left(\dfrac{d^2X}{dt^2}\right) T^2}{\left(\dfrac{dX}{dt}\right)\left(\dfrac{dT}{dt}\right)} = \frac{-n\left(\dfrac{dX}{dt}\right)(T^2)}{(1 - X)\left(\dfrac{dT}{dt}\right)} + \left(\frac{E}{R}\right) \tag{3.41}$$

This equation can be used to determine the reaction order and activation energy.

### 3.2.9   Varhegyi's Integral Solution

Varhegyi (1978) has provided the integral of the rate expression considering the temperature dependency of the pre-exponential factor. He considered the rate expression of the form

$$k = AT^b \exp\left(-\frac{E}{RT}\right) \tag{3.42}$$

and then provided the integral of equation (3.42) with respect to temperature

$$\int k \, dT = \int AT^b \exp\left(-\frac{E}{RT}\right) dT \tag{3.43}$$

Substituting $y = E/RT$ and $b = s - 2$ in equation (3.43)

$$\int T^b \exp\left(-\frac{E}{RT}\right) dT = p_s(y) = \left(\frac{E}{R}\right)^{(s-1)} \int y^{-s} \exp(-y)\, dy$$

$$(3.44)$$

The solution of the integral is provided by using continued fractions due to Legendre

$$p_s(y) = \frac{y^{1-s}e^y}{(y + s/(1 + 1/(y + (s + 1)/(1 + 2/(y + (s + 2)/(1 + \cdots))))))}$$

$$(3.45)$$

If equation (3.45) is truncated at the sixth sign of division, the relative error of the approximation is about $10^{-5}$ at $y$ values of 10. In thermal analysis, the values of $y$ are not less than 10.

Another solution to the $p_s(y)$ integral with relative error being of the same order of magnitude as in equation (3.45) is

$$p_s(y) = y^{-s}e^{-y}\left(1 + \left(\frac{a_1}{(y + 1)}\right) + \left(\frac{a_2}{(y + 1)(y + 2)}\right)\right.$$
$$\left. + \cdots + \left(\frac{a_n}{(y + 1)(y + 2)\cdots(y + n)}\right)\right)$$

$$(3.46)$$

This series was proposed by Schlomlich (Van Krevelen et al., 1951). The method for the determination of coefficients $a_1 \cdots a_n$ is given by Bateman and Erdelyi (1953). Varhegyi also lists Pede's approximation:

$$ps(y) = y^s e^y \left[\frac{(y + 1)}{(y + s + 1)}\right]$$

$$(3.47)$$

The rate equation for decomposition can be written as:

$$\int \frac{dX}{f(X)} = g(X) = \left(\frac{1}{a}\right)\int [k(t)\, dT]$$

$$(3.48)$$

Using equation (3.44), equation (3.48) can be written as

$$g(X) = \frac{A}{a}\left(\frac{E}{R}\right)^{s-1} p_s(y)$$

$$(3.49)$$

but $p_s(y)$ is a product of $y^{-s}e^{-y}q_s(y)$, where $q_s(y)$ is the approximation for last term in equation (3.47). Substituting this value for $p_s(y)$, we can write

$$g(X) = \left(\frac{AR}{aE}\right) T^s e^{-y} q_s(y)$$

$$(3.50)$$

or

$$\ln \frac{g(X)}{T^s} = \ln \left( \frac{AR}{aE} \right) + \ln q_s - y \tag{3.51}$$

The term $\ln(q_s)$ can be approximated by expansion using Taylor's series

$$\ln(q_s) = c_0 + c_1 y \tag{3.52}$$

The values for $E$ and $A$ are evaluated in the following manner: At first the term $\ln[g(X)/T^s]$ is obtained from the experimental data. Then these values are approximated by a linear function of $1/T$:

$$\ln \left[ \frac{g(X)}{T^s} \right] = B_0 + B_1 \left( \frac{1}{T} \right) \tag{3.53}$$

At the first approximation we can get an estimate of coefficient $B_1$, which is equal to $-E/R$. Then an average temperature $T$ is chosen somewhere in the middle of the temperature interval of decomposition; then using the first approximation for $E$, the corresponding value for $y = E/RT$ is determined. Subsequently, using this value of $y$, the coefficients for the expansion of $\ln(q_s) c_0$ and $c_1$ are calculated:

$$c_1 = \frac{s}{[(y + 1)(y + s + 1)]} \tag{3.54}$$

$$c_0 = \ln \left[ \frac{(y + 1)}{(y + s + 1)} \right] - c_1 y \tag{3.55}$$

Knowing the values of $c_1$ and $c_0$ will help in determining the more correct values for $E$ and $A$ by using equations (3.49)–(3.51) and rearranging them to get

$$B_1 = -(1 - c_1) \frac{E}{R} \tag{3.56}$$

$$B_0 = \ln \left( \frac{AR}{aE} \right) + c_0 \tag{3.57}$$

The difference in the prediction of the $E$ value or the activation energy by going through this method is only of the order of 3–5%, but the value for $A$ is improved by a factor of 50%.

Flynn and Wall (1966) and Satava and Skavara (1978) have also shown in their independent studies that the linear dependency of pre-exponential factor on the temperature has no significant effect, but the change in heating rate has a dramatic effect. However, they do no combine this effect with the temperature dependency of $A$ factor with the power law.

## 3.2.10 Gaur and Reed Method

Varhegyi's integral has shown that if the pre-exponential factor is not considered as temperature dependent with power function its estimation can be off by 40–50%. Satava and Skavara (1978) have shown in their independent study that the effect of heating rate and the change in pre-exponential term result in similar changes in the weight loss curve of a sample. Flynn and Wall (1966) have shown that there is negligible change on the pre-exponential factor if it is considered as a linear temperature function, but on the other hand they also state that the heating rate makes a lateral shift on the curve.

Gaur and Reed (1994) have developed an equation that incorporates the change in pre-exponential factor as a function of heating rate.

The rate of decomposition can be given as

$$\frac{dX}{dt} = A_0 \exp\left(-\frac{E_0}{RT}\right) f(X)^n \tag{3.58}$$

where

$$A_0 = k'T^m \tag{3.59}$$

$$E_0 = E_{\text{exp}} - RT \tag{3.60}$$

In the above equations, parameters $k'$ and $E_0$ are constants that are independent of temperature, while $A_0$ and $E_{\text{exp}}$ are temperature dependent. Equations (3.59) and (3.60) are the modified forms of the original Arrhenius derivation, where the exponent $m$ has an explicit value of 0.5. Since $A_0$ varies with temperature, the equality of the temperature independent term $k'$ at two different temperatures can be written as

$$k' = \frac{A_0}{T^m} = A_{0,i} = T_i^m \tag{3.61}$$

or

$$A_{0,i} = A_0 \left(\frac{T_i}{T}\right)^m \tag{3.62}$$

where

$T$ refers to the condition at which kinetic parameters are determined

$T_i$ refers to the temperature at which the predictions for solid decompositions are to be made

Substitution of equation (3.62) into equation (3.58) at temperature $T_i$ gives

$$\frac{dX}{dt} = A_0 \left(\frac{T_i}{T}\right)^m \exp\left(\frac{E_0}{RT}\right) f(X)^n \tag{3.63}$$

The temperature ratio in the above expression can be written in terms of heating rates $(a_i/a)^m$, where $a_i = dT_i/dt$ and $a = dT/dt$. Since rate constant for this reaction is close to zero at $T < T_0$, $T_0$ can be considered as zero degrees centigrade. For limits $T = 0$ at $t = 0$ and $T = T$ at $t = t$ and at equal time intervals

$$\frac{a_i}{a} = \frac{T_i}{T} \tag{3.64}$$

Substituting (3.64) in (3.63) gives

$$\frac{dX}{dt} = A_0 \left(\frac{a_i}{a}\right)^m \exp\left(-\frac{E_0}{RT_i}\right) f(X)^n \tag{3.65}$$

Expressing the above rate expression for constant heating rate TG data, we get

$$\frac{dX}{dT} = \frac{A_0}{a_i} \left(\frac{a_i}{a}\right)^m \exp\left(-\frac{E_0}{RT_i}\right) f(X)^n \tag{3.66}$$

The solution for this equation has been provided by the authors for $f(X)^n = (1 - X)^n$ function using Doyle's approximation for the integration.

For $n = 1$

$$-\ln(1 - X) = \left(\frac{E_0 A_0}{Ra_i}\right) \left(\frac{a_i}{a}\right)^m 10^{(-2.315-0.4567E_0/RT)} \tag{3.67}$$

and for $n \neq 1$

$$\frac{[1 - (1 - X)1 - n]}{(1 - n)} = \left(\frac{E_0 A_0}{Ra_i}\right) \left(\frac{a_i}{a}\right)^m 10^{(-2.315-0.4567E_0/RTi)} \tag{3.68}$$

## 3.3  DETERMINATION OF KINETIC PARAMETERS USING DIFFERENTIAL THERMAL ANALYSIS

Several methods have been proposed for the determination of reaction kinetics by using a differential thermal analysis curve. Some of the commonly used methods are the Kissinger method, the Tateno method, and the method of Borchardt and Daniels (1956, 1957). In the Kissinger (1957) method, the variation in peak temperature with heating rate has been used to determine the activation energy. The Tateno (1966) method is described by the means of transfer functions and gives values for reaction order and activation energy.

In this book we present the method of Borchardt and Daniels, which is the most extensively used method. This method was developed for homogeneous reactions.

The application of DTA studies to determine the kinetic parameters was successfully attempted by Borchardt in his doctoral work at the University of Wisconsin (1956). In his work, he determined the reaction kinetics by using only one nonisothermal curve. Based on this work, Borchardt and Daniels (1957) developed the method for the determination of reaction kinetics for homogeneous solutions. This method employs the development of a heat balance equation around the sample:

$$C_p \, dT = dH + k\Delta T \, dt \qquad (3.69)$$

where

$dH$ = heat of reaction
$C_p \, dT$ = enthalpy of the reaction solution
$k\Delta T \, dt$ = heat transferred into the cell from the surroundings

$$dH = C_p \, dT + k\Delta T \, dt \qquad (3.70)$$

If $C_p$ and $k$ are independent of temperature, then for $t_0$ to $t_\infty$:

$$\Delta H = C_p(\Delta T_\infty - \Delta T_0) + k \int \Delta T \, dt \qquad (3.71)$$

At $T = \infty$ and $T = 0$, $\Delta T_\infty$ and $\Delta T_0$ are zero as $\Delta T$ is the temperature difference between the sample and the furnace. Therefore

$$\Delta H = kA \qquad (3.72)$$

where $A$ = total area under the curve. Since $dH$ is the heat transferred due to reaction on per mole basis, $-kA/n_0$, we can write the following expression for fractional conversion in homogeneous reaction as

$$dH = -kA/n_0(dn) \qquad (3.73)$$

Substituting (3.73) into (3.69) gives

$$-\frac{dn}{dt} = \left(\frac{n_0}{kA}\right)\left[C_p\left(\frac{dT}{dt}\right) + k\Delta T\right] \qquad (3.74)$$

where number of moles present at any instant are given as

$$n = n_0 - \int \frac{dn}{dt} \qquad (3.75)$$

The development of the kinetics method for heterogeneous systems was proposed by Blumberg (1959). This method is an extension of the method presented by Borchardt and Daniels for the homogeneous reactions. However, this extension has been subject to lot of criticism due to its assumption of negligible temperature gradient within the solid sample. In addition, the relationship of the peak area in a thermogram to the heat of reaction has also been reported by Ramachandran and Bhattacharya (Ramachandran, 1954).

*Other Effects in Understanding DTA Data*

During dynamic heating there is a chance of temperature lag increase with increasing heating rate. The slower the rate of heating, the smaller the variation. However, with a slow rate of heating, the temperature differential between the reference material and the substance examined will be less. This results in rounded peaks that tend to be more significant as the heating rate is lowered. On the other hand, with a rapid heating rate the peak becomes intense but a lot of details are lost. The theory behind this phenomenon is explained as follows: The temperature at any point lags behind that of the holder by an amount which depends only on the position of the point. This quasi steady state is reached after a time $\tau$ and for a cylindrical sample of radius $r$ and length $4r$.

$$\tau = \frac{1.08\rho C r^2}{\lambda} \tag{3.76}$$

Changes in thermal conductivity of the sample may result from temperature variations and from changes in composition produced by the reaction. The heating rate has an effect on peak height and peak width. A slow heating rate requires a sensitive recording system, while with fast heating rate the neighboring peaks tend to coalesce. The optimum heating rate depends on the nature and characteristics of the sample and reference material. The International Geological Congress, London, recommends a heating rate of 10°C/min. Under these conditions peaks are of satisfactory size, overlap of neighboring peaks is not excessive, and the time for determination is reasonable. The thermal effects in the sample are not uniform during the period of reaction. These can be reduced by using small samples and slow heating rates. The effect of particle size has been considered by several investigators. Norton (1939) has pointed out that finer particles give up their heat more rapidly.

# 4

# EXPERIMENTAL APPARATUS AND DATA COLLECTION

## 4.1 SEIKO TG/DTA

The data reported in this book were collected using a Seiko SSC 5200 TG/DTA thermal analysis apparatus. The schematic of the instrument is shown in Figure 4.1. This apparatus has two horizontal balance arms containing thermocouples to support the pans and the sample. The close contact of thermocouples with the sample allows for very accurate measurement of the sample temperature (within 3°C), with negligible deviation over the entire heating range of the furnace. The two balance arms permit taking simultaneous TG/DTA data on the same sample and minimizes the errors in analysis caused by the sample heterogeneity. One balance arm is used to support the reference inert material and the other arm supports the sample under study. The relative weight change between the two sample pans as a function of temperature at a given heating rate give data for thermogravimetric analysis and the change in energy requirements as the reaction progresses is measured in terms of micro volts to provide data for differential thermal analysis.

The gas flows parallel to the balance arm and across the sample pan instead of parallel to the sample pan, as is the case with most other models. This arrangement minimizes buoyancy effects caused due to change in gas density with change in temperature. We found that the deviation in the sample weight from ambient to 1100°C

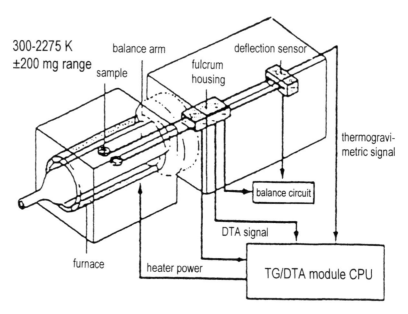

**FIGURE 4.1** Schematics of Seiko SSC 5200 TG/DTA.

was about 30 μg for 5 mg sample, resulting in about 0.5% error in weight measurements. In our experience, this is fairly negligible in comparison to most other thermal balances. The maximum permissible sample weight limit is 200 mg.

This TG/DTA model permits introduction of two reaction gases in any proportion at a flow rate of 100–600 cc/min. The switching of gases is done with the help of a gas flow controller, which is regulated by the software provided along with the unit. In our opinion this software is user friendly and versatile.

One of the drawbacks of the system is that the software is not MS-DOS compatible and has its own operating system. This slows the transfer of data to other MS-DOS compatible programs for further analysis, such as for the determination of kinetic parameters. In our study, this problem was overcome by digitizing the data or transferring it by using data conversion software.

## 4.2   PROCEDURE FOR DATA COLLECTION

The samples were at first powdered by milling or grinding and heated to 120°C at 10°C/min, and were held there isothermally to allow for the removal of physical moisture. Subsequently, the sample temperature was raised at 10°C/min to 950°C in nitrogen atmosphere at a flow rate of 300 cc/min. The net weight loss between 120° and 950°C was recorded as *total volatile matter*. The residual char was then burnt in air at 950°C until a constant weight was obtained. The weight loss due to this combustion is reported as *fixed carbon*. The weight remaining is reported as the *ash content* of the sample. The volatile portion of the fuel was divided into two parts. The first part includes volatiles that are liberated between 120° and 650°C. This accounts for the major portion of the volatile fraction in biomass samples and is largely responsible for flaming combustion. This volatile fraction is sometime referred to as *primary volatiles*. The second portion of the volatile gases, sometimes termed *secondary volatiles*, is liberated between 650° and 900°C. This volatile fraction takes little part in flaming pyrolysis and does not inhibit surface combustion. Unless stated otherwise, all of the analysis was carried out with samples having particle size in the range of +10 to −100 mesh. Typical thermal analysis data for cellulose sample as obtained from Seiko TG/DTA unit are shown in Figure 4.2. It can be seen that there are three curves as a function of temperature. The TG curve shows the percentage of sample reacted as a function of temperature, the DTG curve gives the percentage weight loss/°C of temperature increase, and the DTA curve gives the energy change in microvolts as a function of reaction temperature.

*Thermogravimetric analysis*: In TG analysis, the weight of a sample is recorded as a function of time or temperature while heating the sample in a preset time-temperature program. A typical

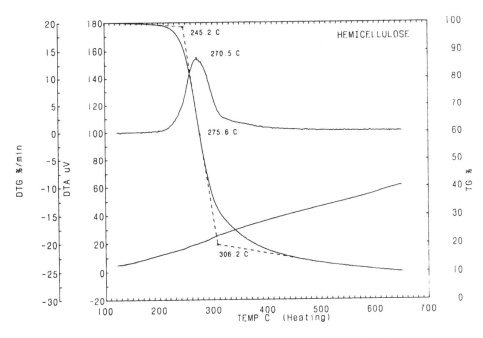

**FIGURE 4.2** A typical TG-DTG-DTA thermogram in this book.

thermogram showing weight as a function of both time and temperature is shown as curve $a$ in Figure 4.2 for the cellulose sample heated at 10°C/min in a flowing nitrogen atmosphere. The temperature at which there is a first sign of weight loss in the sample is shown at $T_{initial}$. $T_{start\text{-}point}$ is shown as the beginning of the $S$ curve as obtained from the method for determining glass transition temperature, and $T_{end\text{-}point}$ in the temperature at the end of the $S$ curve. $T_{mid\text{-}point}$ refers to the midpoint between the beginning and the end of $S$ curve.

*Differential thermogravimetric analysis curve*: It is useful to differentiate TG data with respect to time or temperature in order to show the rate of weight loss. The curve thus obtained is called a DTG curve and is shown as curve $b$ in Figure 4.2. Seiko software provides this information as one of the standard outputs. DTG helps in distinguishing between the various reaction steps taking place over the entire temperature range and in identifying the rate of reaction for each reaction step as a function of temperature. The point $T_{max}$ on this curve shows the temperature at which the maximum rate of reaction occurs.

*Differential thermal analysis*: In DTA, the temperature difference between the sample and a standard inert material are recorded as a function of temperature, thus showing the occurrence of either endothermic or exothermic reaction or phase change. The first inflection point signifies the beginning of the transition and the second inflection point shows the end. The area within the peak gives an estimate for heat of reaction for that particular transition. The DTA curve obtained from Seiko TG/DTA is shown curve $c$ at the bottom of Figure 4.2.

Almost all of the data given in this book are presented with the same scale coordinates so that the data for different samples can be easily compared.

Another widely used tool is the *proximate analysis*. The proximate analysis (ASTM, D3172-75) gives the moisture content of the sample, the amount of material that volatilizes up to 950°C in an inert atmosphere, the fixed carbon, and the ash content. Except for moisture, the other properties are listed for most samples in the table accompanying each diagram.

## 4.3  CALIBRATION OF TG/DTA

Thermal analysis is not an absolute measuring technique and hence calibration is of prime importance. The calibration of the sample temperature to the thermocouple reading was performed by determining the curie point for some ferromagnetic samples in the temperature range of 250–800°C. In this procedure, a magnet was installed just outside the furnace but close to the sample pan supports. Subsequently, the ferromagnetic samples whose curie point temperature (the temperature at which these samples lose their magnetic property) was known were placed in the sample pan. The TG/DTA furnace was then heated from ambient to 900°C at designated heating rate. The temperature at which these samples lose their magnetism is signified by the weight change due to the absence of magnetic force caused by the magnet placed outside the furnace. The deviation between the furnace temperature and the temperature at which the sample is supposed to lose its magnetic property is the measure of error between the sample temperature and the one that is being recorded by the furnace. ICTA has come up with some recommended ferromagnetic samples for the calibration of TG/DTA instruments (Wunderlich, 1990). We conducted this calibration procedure over the entire range of experimental temperatures at various heating rates. The maximum deviation was within ±3°C. Once a calibration for a temperature has been performed, then most of the thermobalances are capable of providing very precise data for long periods of time if the system is maintained properly, i.e., there is no contamination of the sample crucible, thermocouple, or furnace wall, and, most importantly, the balance mechanism has not been disturbed. However, it is important to perform these calibrations periodically.

## 4.4   COLLECTION OF SAMPLES

The samples reported in this book were collected from a wide variety of sources specializing in various aspects of biomass. The sources are identified in the appendix. In general the samples were reduced to +10 to −100 mesh by milling or grinding. Wherever it was necessary to obtain an average of a large specimen, whole plants were dried and milled. These samples are stored in vials and can be made available to interested parties for the sake of comparative testing.

## 4.5   ERRORS IN THERMAL ANALYSIS

In TG data collection, many aspects must be examined before the data can be said to be the true representation of the decomposition kinetics or for that matter any other gas-solid reaction. The common sources of error are as follows:

1. The presence of temperature gradients in the sample.
2. Use of an excessive heating rate.
3. The existence of a partial pressure of the product gases over the sample.
4. The effect of impurities and mineral content.
5. Buoyancy effects.

We will discuss some of the effects due to these conditions in the TG/DTA data collection.

## 4.5.1 The Presence of Temperature Gradients in the Sample

If there is a temperature gradient within the sample from the surface to the center, then the TG data collected for the decomposition cannot provide the correct decomposition kinetics because the sample temperature is not defined. This aspect limits the size of the sample and the heating rate to which it should be exposed. One way to correct for this error would be to go for lower heating rates with the increasing sample size. However, the isothermality of the sample can be defined in terms of a nondimensional quantity called the Biot Number ($N_{Bi}$). It is the ratio of the external heat transfer to the surface of the sample to the internal heat transfer within the sample by thermal conduction:

$$N_{Bi} = \frac{h}{\kappa r}$$

where

$h$ = heat transfer coefficient to the sample surface
$\kappa$ = thermal conductivity
$r$ = a characteristic dimension (the minimal diameter) of the sample

The higher the value of $N_{Bi}$, the greater the temperature gradient within the sample. It has been found experimentally that temperature gradients become negligible (<5%) at $N_{Bi} < 0.1$ (Wilty et al., 1985).

## 4.5.2 Use of an Excessive Heating Rate

Most biomass decomposition is either a simple decomposition of components such as cellulose, hemicellulose, or lignin or the decomposition of mixtures of these. In the case of some biomass species, the decomposition of individual components tends to overlap, which makes it difficult to distinguish one from the other. This is also true for some polymeric materials, like polyvinylchloride (PVC) found in municipal solid waste. Polyvinylchloride has two or three reactions in a series. If the heating rate of the sample is very high, say 100°C/min, then these reactions tend to overlap on the thermograms and one would then tend to interpret this as a single step decomposition. This leads to an erroneous determination of kinetic parameters. It is, therefore, advisable to conduct experiments for decomposition kinetics under slow heating rates. The London Geological Congress recommends the use of 10°C/min. It has been found that this heating rate works well for most of the biomass samples. However, one's own judgment will ultimately determine what heating rate is best. For complex reactions such as a mixture of two of three species, it is advisable to test samples at heating rates as low as 1°C/min. The International Conference on Thermal Analysis (ICTA) recommends the reporting of heating rate along with every thermogram.

### 4.5.3 The Existence of Partial Pressure of the Product Gases Over the Sample

For reactions that are reversible in nature, such as the decomposition of $CaCO_3$ to $CaO$ and $CO_2$, a partial pressure of the product gases would cause the slowing of the forward kinetics (Varhegyi et al., 1988). It is therefore important to keep the concentration of the product gases at a value where the kinetics of the reversible reaction has a negligible effect on the decomposition process. This can be accomplished by keeping the flow rate of sweep gas sufficiently high. Care should be taken in doing this because, at high flow rates of the sweep gas, the errors due to the buoyancy effect can become dominant and increase the contribution of systematic error on the data collection. For reactions that are not reversible in nature, like the depolymerization of PVC, one would think that low flow rates of sweep gas will not contribute to significant error. However, this thinking may lead to some problems because the vapor pressure of the product gases in some cases can have a catalytic effect on the decomposition process.

### 4.5.4 The Effect of Impurities and Mineral Content

Impurities such as ash mineral matter have an effect on the kinetics of decomposition. Therefore, to determine the true kinetics of a decomposition, it is advisable to remove the ash content by any method such as acid wash followed by the neutralization with an alkali (Varhegyi et al., 1993). However, it is true that the naturally occurring mineral matter would in most cases be present in the actual application, so in this case one should determine the kinetics in the presence of the mineral matter. In such cases it is advisable to report the source and treatment of the sample. It is well known that samples of the same species of biomass samples may have different inorganic contents, depending upon the geographic location from which they come from. This change is largely due to the geological form for forestry plants and the change in the fertilizer type for the agricultural crops. The analysis presented in this book is without doing any pretreatment to the sample and, therefore, wherever possible the source of the sample has been listed.

## 4.5.5 Buoyancy Effects

Another important source of error in TG data collection relates to the buoyancy effects due to change in gas density at increasing temperatures. The mass change in the gas can be calculated by the application of the ideal gas law and can be used as a buoyancy correction factor.

## NOTE TO READERS

Because of these various factors that can contribute to errors in TG/DTA data collection, it has been found that the results for the same sample from two laboratories operated under presumably identical conditions can be somewhat different. In view of these differences in the literature, ICTA recommends a procedure for data reporting that we have reproduced in Chapter 3.

# 5

# PYROLYSIS OF THE COMPONENTS OF BIOMASS

Devolatilization forms the major step in any thermochemical conversion process involving biomass materials. This is primarily due to the fact that biomass materials, on a weight basis, comprise about 80% volatile fractions and 20% solid carbonaceous residue. During the *devolatilization* or *pyrolysis* process for biomass materials, heat is applied to biomass particles through a direct or an indirect mechanism under inert or oxygen lean conditions. This process usually takes place in the temperature range of 300–700°C and results in the liberation of volatile hydrocarbons from hemicellulose, cellulose, and lignin. These volatiles are subjected to thermal cracking or combustion depending on the reactor configuration. The pyrolysis process of biomass materials involves several reactions, which makes it a complex reaction phenomenon. Hence, in order to understand the pyrolysis mechanism of biomass particles as a whole, it is best to independently understand the pyrolysis mechanism of individual fractions present in the biomass sample, and then combine them to get an understanding of how a biomass sample as a whole undergoes the transformation. In this chapter, the discussion on pyrolysis of biomass components is presented along with several thermograms to give a general idea of thermal degradation of these components in terms of temperature range, kinetic behavior, and catalytic conversion.

Some aspects of the basic kinetic theories applicable to devolatilization process, such as the Arrhenius rate law, temperature dependency on the pre-exponential term, the compensation effect, nonisothermal kinetics, and thermogravimetric methods, are presented in Chapter 1. For details, the reader is advised to refer to the standard kinetic books (see Appendix C).

Biomass samples generally comprise about 80% volatile matter on a weight basis, the balance of 20% being char and ash. Since volatile products form a large fraction of any biomass sample, in contrast to a coal sample, they play a significant role in the pyrolysis of biomass materials, product gas composition, and the extent of char formation.

It is well known that biomass components degrade over a wide temperature range and result in different characteristics based on the heating rates to which they are subjected. In this chapter, we have attempted to clarify their degradation on the temperature scale and the heating rate. The primary reason for taking this approach lies with the fact that, in general practice, there are two types of biomass conversion systems. In one, the conversion of biomass takes place at slow heating rates; systems such as charcoal making and fixed bed gasifiers can be placed in this category. In the second category are processes in which biomass materials are converted at high heating rates; systems such as liquefaction processes and fluidized bed technologies can be placed in this class.

In our attempt to understand the kinetics of biomass pyrolysis, we have differentiated these two classes because they result in different apparent kinetic parameters. It is our belief that by incorporating the necessary temperature correction factors discussed in Chapters 3 and 12 to the data presented in this book, researchers and engineers will be able to generate the kinetic parameters useful to their specific processes.

## 5.1  HEMICELLULOSE

*Hemicelluloses* are a mixture of polysaccharides that form an integral part of the cell wall. There are three important forms of hemicelluloses:xylans, glucomannans, and arabinogalactans. The basic structure of *xylans* found in plants is a linear backbone of 1,4-anhydro-D-xylopyranose units. Arabinoxylans are found primarily in grains such as wheat and corn and in grasses such as esparto. One of the simplest xylans is from esparto grass, which contains only a small fraction free of any extraneous sugar unit. Xylans free of extraneous sugars but combined with uronic acids are found in hardwoods. The degree of polymerization reported for the bulk of the xylan-uronic acid system from hardwoods is around 150–250. A number of xylan fractions have been reported to contain arabinose and uronic acids. In wheat straw, these have been fractionated into arabinose-rich and arabinose-free polysaccharides.

*Arabinogalactans* have been found as water-soluble polysaccharides and the physical and chemical properties of these substances vary between species. The arabinogalactans are highly branched structures. Methylation and hydrolysis of these compounds yield fractions like 2,4 dimethyl-D-galactose, 2,3,4 trimethyl D-galactose, and 2,3,4,6 tetramethyl D-galactose.

The glucomannans found in wood are of relatively small molecular weight. The average degree of polymerization varies from 100 to 200. The ratio of glucose to mammose in these polysaccharides varies from 1:1 to 1:4, and the structure is primarily linear chains. The glucomannans are considered to be copolymers in which the length of the sequences of a single sugar depend on the ratio of the monomer.

The distribution of hemicellulose in plant tissues varies with different species. The weight ratio of the individual sugars in softwoods is found to be: glucose 61–65, mannose 7–16, galactose 6–17, xylose 9–13, and arabinose less than 3. In hardwoods, the similar analysis shows: glucose 55–73, xylose 20–39, galactose 1–4, and mannose 0.5–4. The fractions present in hardwoods are more soluble in comparison to softwoods. The soluble fractions of hardwood contain glucose, xylose, mannose, and galactose, while those of softwood are arabinoglucoronoxylan. Table 5.1 gives a qualitative idea of hemicellulosic polymers and their distribution in conifer and deciduous trees.

**TABLE 5.1**

Hemicellulosic Polymers and Their Distribution in Deciduous and Conifer Trees

| Polymer | Deciduous | Conifer |
|---|---|---|
| 4-O methylglucuronoxylan | 80–90 | 5–15 |
| 4-O methylglucuronoarabinoxylan | <1 | 15–30 |
| Glucomannan | 1–5 | 60–70 |
| Galactoglucomannan | <1 | 1–5 |
| Arbinogalactan | <1 | 15–30 |
| Other galactose polysaccharides | <1 | <1 |
| Pectin | 1–5 | 1–5 |

## 5.1.1  Pyrolysis of Hemicellulose at Low Heating Rates

Hemicelluloses are low-temperature volatile substances. At low heating rates, their thermal degradation initiates at temperatures below 200°C, and by 350°C most of the hemicellulose undergoes the devolatilization process. Its degradation process is very similar to cellulose pyrolysis, which is more widely known to readers of this field and is discussed in the next section of this chapter. A typical non-isothermal pyrolysis thermogram for a hemicellulose sample is given in Fig. 5.1

At about 200°C, hemicellulose undergoes rapid decomposition, producing a variety of anhydrosugars, light volatile organic compounds, permanent gases like $H_2O$, $CO_2$, and CO, and char. The anhydrosugars in a demineralized sample (ash free) and at atmospheric pressure transform into vapor phase with little or no char formation due to the absence of catalytic action of mineral matter; however, in the presence in ash they rapidly decompose to form a substantial part of solid char (5–10%). From the gasification and combustion point of view, this would mean that for biomass materials with high ash content, such as rice hulls, hemicellulose would be responsible for significant contribution to char conversion step. Sufficient proof to this effect has been provided by the works of Shafizadeh and coworkers (1973, 1979, 1980), which show that compounds like ZnCl catalyze the exothermic decomposition of levoglucosan, a product of hemicellulose to produce char, $H_2O$, and $CO_2$

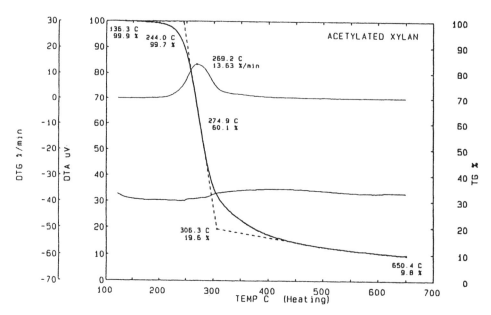

**FIGURE 5.1**  A typical nonisothermal hemicellulose pyrolysis thermogram.

between 120 to 180°C. Similar studies conducted by Broido (1976) indicates that Lewis acids are responsible for converting hemicellulose to char.

Another important contribution during the hemicellulose pyrolysis is made by the degradation of mono- and polysaccharide fractions. Hounimer and Patai (1967) found that they are the dominant contributors to CO and $CO_2$ between 175 and 275°C. The major contribution is made from the $C_1$ atom of the hydrocarbons, which is also supplemented by the contribution from $C_2$ carbon fractions. The authors hypothesize that the differences in the amounts of products formed during this pyrolysis regime is due to the presence of mineral matter. However, they were not able to suggest which component of the ash (mineral matter) is largely responsible for the high liberation of permanent gases like CO and $CO_2$ along with the formation of char. In addition, it was also not clear which bond cleavages occurred to result in the formation of permanent gases and solid char.

To better understand the degradation of these compounds, efforts were made to study the decomposition behavior of model compounds like D-glucose that more closely resemble the properties of mono- and disaccharides. It was found that the product gases primarily contained organic volatile products with increasing temperature up to 275°C, and there was no evidence for the formation of char CO and $CO_2$. To resolve this perplexing situation, experiments were conducted with other model compounds such as D-erythrose, D-xylose, D-ribiose, D-arabinose, D-glucurono-6,3-lactone, cebobiose, maltose, lactose, and sucrose. It was concluded that by dehydration, degradation, and condensation reactions, all of these carbohydrates formed similar polymeric intermediates and little or no fractions of CO and $CO_2$ were traced. These polymeric intermediates undergo secondary thermal degradation, which results in the formation of CO and $CO_2$ (Heyns and Klier, 1968). This shows that the release of permanent gases and the resultant char formation is not due to the primary volatile reactions of hemicellulosic matter but due to secondary thermal cracking and catalytic degradation by mineral matter of the volatile products. This is in agreement with the observations reported earlier with a hemicellulose sample from biomass material. A similarity of results between previous studies and this study suggests that the increase of char formation is directly related to the ash percentage and its composition in the biomass material. To support the concept of increased char formation during hemicellulose pyrolysis due to the presence of mineral matter, Furneaux and Shafizadeh (1979) conducted studies with an acid-pretreated sample. They at first reduced the ash content to

0.1%, a negligible value, and then subjected the sample to thermal treatment in an inert atmosphere. It was found that the pyrolytic char yield was reduced from 33% for an untreated sample to 27% for a demineralized sample at 500°C.

To further understand the mechanism of hemicellulose pyrolysis, Bryce and Greenwood (1963) conducted a series of experiments. They reported that the volatile products of potato starch, a polysaccharide, are qualitatively identical to those evolved from various starches, amylose and amylopectin components, and simple sugars such as D-glucose, maltose, and maltotriose. Primary products of starch pyrolysis consist of levoglucosan, along with a variety of permanent gases, and char. This is in agreement with the work of Heyns and Klier (1968) conducted on mono- and disaccharides, and confirms to certain extent that *polysaccharides from any source produce similar products and hence the results and interpretation can be applied to general degradation of hemicellulose pyrolysis.*

Further work in this regard was done by Shafizadeh and Lai (1973). They conducted a series of experiments involving the vapor-phase of synthetic levoglucosan labeled at the $C_1$, $C_2$, and $C_6$ carbon positions. By tracing the labeled carbon atoms of the major pyrolysis products back to their original positions in the anhydrosugar molecule, the authors were able to deduce the pyrolysis pathways to certain extent. The results indicate that the major products are formed by a variety of pathways. Carbon dioxide and carbon monoxide are primarily derivatives of $C_1$ and $C_2$ carbon atoms. The observed formation of $CO_2$ from the $C_2$ and $C_6$ carbon atoms indicates that extraordinary rearrangements must occur during vapor-phase pyrolysis. The results indicate that the uncatalyzed pyrolysis of levoglucosan more closely resembles the acid-catalyzed than the alkaline-catalyzed results.

The detailed mechanisms of anhydrosugar formation and polymerization from hemicellulose pyrolysis is given by Cemy and Stanek (1977). In their article, the authors note that the radical polymerization of levoglucosan, a major step in hemicellulose pyrolysis, does not take place. Instead, competing reactions that form oligo- and polysaccharides by polymerization are observed up to 250°C. The overall activation energy for polysaccharide thermal decomposition process for char and permanent gas formation is given in the neighborhood of $18-26$ kcal/g mole and the pre-exponential is in the range of $1 \times 10E^4$ to $1 \times 10E^6$ sec$^{-1}$. The reaction process is found to satisfy a first-order rate as per the Arrhenius rate law.

The formation of the polysaccharide and its decomposition in the range of $250-350$°C is given by the following reaction mechanism as per Shafizadeh and his colleagues (1976):

$$G - F \rightarrow G + F^* \qquad \text{(rate determining step)}$$
$$G - F + F^* \rightarrow F - G - F$$

where G and F are the glucose and fructose constituents of sucrose. Richards and Shafizadeh (1978) note that no gas was evolved at 194°C during the formation of polysaccharides and that the polysaccharide decomposed upon further heating to 300°C.

Hounimer and Patai (1967) identified the main source of anhydrosugar formation to be depolymerization of the glucan created from the D-glucose at lower temperatures. Concurrent with the dehydration reactions was the fragmentation of the carbon skeleton, leading to the formation of furans and carbonyl compounds. The fragmentation reactions are facilitated by the intramolecular dehydration reactions. The rate of the fragmentation reactions was observed to increase more rapidly with increasing temperature than did the rate of the dehydration reactions.

The permanent gases, CO and $CO_2$, as well as $H_2O$ and various anhydrosugars are the major volatile products of starch pyrolysis in the temperature range of 275–350°C. The formation of these products takes place around 300°C largely by following first-order reaction kinetics. The combined decomposition reactions of starch, amylose, amylopectin, and cellulose in this temperature range are found to have an apparent activation energy lying between 26 and 30 kcal/g mole. The pre-exponential term for these reactions is in the order of $1 \times 10E^4$ $sec^{-1}$ (Puddington, 1948; Bryce and Greenwood, 1966). Because the production versus time curves for $CO_2$, CO, and $H_2O$ were all nonsigmoidal, that is, the curves manifested a continuously decreasing first derivative, it was concluded that there was no induction period, autocatalysis, or liquid phase present during the decomposition of starches. However, no accurate measurements have been made to this effect.

### 5.1.2 Overall Kinetic Parameters for Hemicelluloses

The overall kinetic parameters for hemicellulose degradation as reported by Chornet and Roy (1980) is around 29 kcal/mole, while the one reported by Williams and Besler (1994) varies from 61 kcal/mole with a pre-exponential factor of $1.86 \times 10^{22}$ $sec^{-1}$ at 5°C/min heating rate to 29 kcal/mole with pre exponential factor of $1.61 \times 10^9$ $sec^{-1}$ at 80°C/min heating rate.

### 5.1.3 Hemicellulose Pyrolysis at Rapid Heating Rates

Hemicellulose pyrolysis at high temperatures is common under high heating rates and involves both solid- and gas-phase chemistry. This type of hemicellulose pyrolysis is typical in a fluidized-bed gasifier or an entrained-bed gasifier.

When a hemicellulose sample is subjected to a very rapid heating, it undergoes pyrolysis at temperatures exceeding 500°C. The shift in the temperature for thermal degradation takes place due to temperature effects on pre-exponential term of the kinetic equation, as discussed in chapter 1. In these situations, the hemicellulose degradation results in increased amount of volatile products and char formation is of reduced order; therefore, under the rapid heating mode, pyrolysis of biomass plays an even more critical role than it does at low heating rates. Although it is believed that all biomass materials can be completely volatilized by sufficiently rapid heating, it has been found that for some samples, such as the Douglas fir, a small amount of char (less than 5%) is still formed.

The high-temperature, vapor-phase pyrolysis of volatile matter derived from hemicellulose typically results in the formation of a hydrocarbon-rich synthesis gas and a refractory tar. Under these conditions, intermediate products that are highly unstable at elevated temperatures at low heating rates are observed in significant quantities (Milne, 1981). Quantitative studies of the effects of heating rate and pressure on the pyrolysis of hemicellulose present in the literature suggest that high heating rates favor the transglycosylation and fragmentation reactions, thereby reducing char formation. Conversely, high pressures to a certain extent favor the char forming.

As far as the formation of permanent gases such as CO, $CO_2$, and $H_2O$ are concerned, it is found that the mass ratios for $CO/CO_2$ are less but not negligible at high temperatures (Table 5.2) than

**TABLE 5.2**

Mass Ratios for $CO/CO_2$

| Sample | CO wt.% | $CO_2$ wt.% | $CO/CO_2$ |
|---|---|---|---|
| Dextrose | 3.6 | 24.8 | 0.145 |
| D-cellulose | 4.3 | 22.1 | 0.195 |
| Lignin | 2.1 | 31.6 | 0.066 |
| Kraft paper | 2.4 | 22.8 | 0.105 |
| Leucaena | 3.2 | 49.3 | 0.099 |
| Corn cob | 1.5 | 4.3 | 0.349 |
| Calatropis | 3.1 | 20.8 | 0.149 |
| Newsprint | 1.9 | 20.3 | 0.094 |
| Cow manure | 3.0 | 37.9 | 0.079 |

those obtained at low temperatures (Hopkins et al., 1984). For low-temperature pyrolysis of dextrose, the mass ratios of 0.14 was reported for $CO/CO_2$ by Puddington (1948). In addition, it is found that irrespective of the prevalent gas-phase temperature, a certain amount of CO was always formed. This suggests that a fraction of the carbon atoms composing the volatile matter are dedicated to CO formation by the gas-phase cracking reaction.

Antal's work on the vapor-phase pyrolysis of cellulose-derived volatile matter (1982) has described temperature effects using a semibatch, tubular, laminar flow reactor. Since there is a lot of similarity between hemicellulose and cellulose pyrolysis, we can use some of this work to explain the effects of temperature on hemicellulose pyrolysis.

It is found that (1) the gas-phase cracking of volatiles to permanent gas species has an activation energy, $E$, equal to 49 kcal/gmol; and (2) the gas-phase polymerization of volatiles to a refractory condensable (tar) product has an activation energy of 15 kcal/gmol. Both of these reactions can be described by the first-order rate law. Moreover, gas-phase pyrolysis data at increased temperatures indicate that the molar ratio $CH_4:C_2H_4$ is 1:1 for the vapor-phase pyrolysis reactions. This increased formation of hydrocarbons at higher temperatures seems to come at the expense of CO formation, suggesting the existence of a second low-temperature pathway involving decarboxylation chemistry.

### 5.1.4  Summary of Hemicellulose Pyrolysis

Hemicellulosic materials under low heating rates at atmospheric pressures undergo thermal degradation at temperatures below 200°C. The overall mechanism of hemicellulose pyrolysis under these conditions involves three competing reactions. The first is a low-temperature polymerization process to form components such as polysaccharides, which result in the formation of char, CO, $CO_2$, and $H_2O$. These reactions usually take place at temperatures below 250°C. The kinetics of these reactions can be explained on the basis of a first-order Arrhenius type of equation. The activation energy for this process at low heating rates is in the neighborhood of 18–26 kcal/mole and the pre-exponential term is around $1 \times 10^4$ to $1 \times 10^6$ $sec^{-1}$. At somewhat higher temperatures (250–350°C), these polysaccharides undergo decomposition to generate volatile anhydrosugars and related monomeric compounds. This phase of conversion is also explained by the first-order rate law of the Arrhenius equation with the activation energy in the neighborhood of 26–30 kcal/mole and pre-exponential term in the range of $1 \times 10^4$ $sec^{-1}$.

The third reaction involves the competitive degradation of some xylan type of compounds that also takes place in this temperature range. The fractions are not present in significant quantities. The reaction process for these compounds is much better explained by the Avrami-Erofeyev form of equation, primarily because of very long induction time for these reactions, similar to the nucleation process. However, once the reaction starts, it progresses at a very rapid rate. The kinetic parameters of this type of reaction are not available in the literature. We would therefore advise the reader to look in any standard text book in which Avrami-Erofeyev reaction models are described to calculate the kinetic parameters for this reaction.

The kinetic parameters for initial reactions in a pyrolysis process taking place at rapid heating rates are almost the same as the ones obtained for slow heating rates. However, one has to apply the temperature correction to these kinetic parameters. We recommend that for isothermal conditions this correction be provided as per the derivation given by Laidler (1969). For nonisothermal data, the correction can be provided as per the derivation given by Gaur and Reed (1994). The major difference for rapid-heating pyrolysis as opposed to slow-heating pyrolysis is that the degradation temperature of hemicellulose increases by $50-70°C$ for every order of increases in the heating rate. In addition, in the case of a rapid-heating rate there is much more formation of uncracked volatile fractions in the product gas as compared to slow-heating rates. This is primarily because the kinetic parameters for thermal cracking of volatiles as per the Arrhenius rate law are very severe (E = 49 kcal/mole) in comparison to other reactions, like tar formation (15 kcal/mole). It is important to state here that a certain amount of CO is formed irrespective of temperature conditions even at high heating rates even though its fraction is in reduced order. When subjected to very rapid heating, hemicellulose materials undergo catastrophic fragmentation, producing $H_2$, $CO_2$, and other simple carbonyl compounds.

## 5.2  CELLULOSE

Cellulose is composed of about 50% by weight of most biomass materials and hence it has a major impact on the pyrolysis characteristics of biomass samples. One of the early structure of cellulose was presented by Hibbert (Figure 5.2). The concept of cellulose as a linear macro molecule consisting of anhydro β-glucopyranose was given in the 1920s. It has been found that about 95% of cellulose chains of various lengths are formed of anhydro β-glucopyranosides.

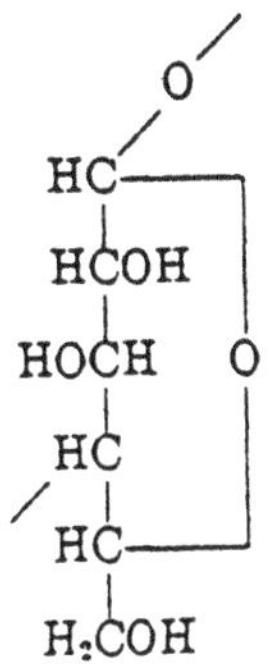

**FIGURE 5.2** Hibbert's model for a cellulose molecule.

These glucosidic bonds could be either an $\alpha$ or $\beta$ glucosidic and could be attached to the 4 or 5 position in the adjacent hexose unit (Figure 5.3). However, due to the large size of the cellulose molecule there are some nonglucosidic bonds, commonly known as weak bonds, that undergo cleavage and help initiate the degradation of the cellulose molecule.

Another unique aspect of the cellulose molecule is that the two terminal groups differ from the glucose residues forming the chain. One of these end groups contains a reducing hemiacetal group that is commonly known as a reducing end group, while the other contains a secondary hydroxyl group and is known as the nonreducing end group (Figure 5.4). These end groups are present in native cellulose and for each glucosidic bond cleavage two new end groups, one of each type, are formed. Due to this fact, end groups in cellulosic molecule are used to determine the molecular weight of cellulose.

The physical structure of cellulose contains a primary wall made up of noncellulosic substances such as waxes. The secondary wall contains almost all of the cellulose (95% by weight) that is present in the fiber. The central canal (lumen) is very narrow and is composed of proteinaceous material. Figure 5.5 gives the schematic illustration of cellulose fiber on a macroscopic level. On the

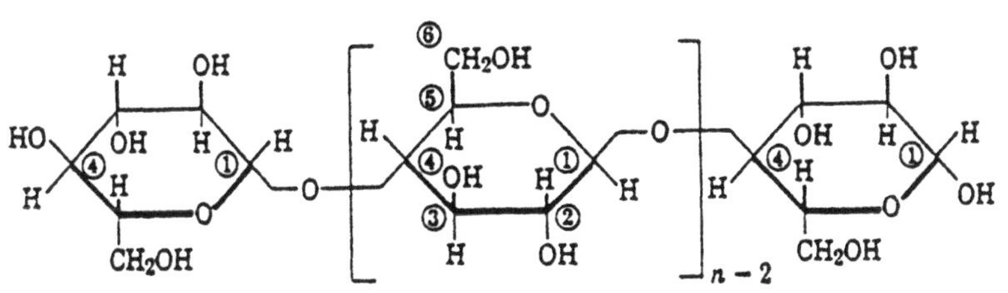

**FIGURE 5.3** Arrangement of glucosidic bonds in a cellulose structure.

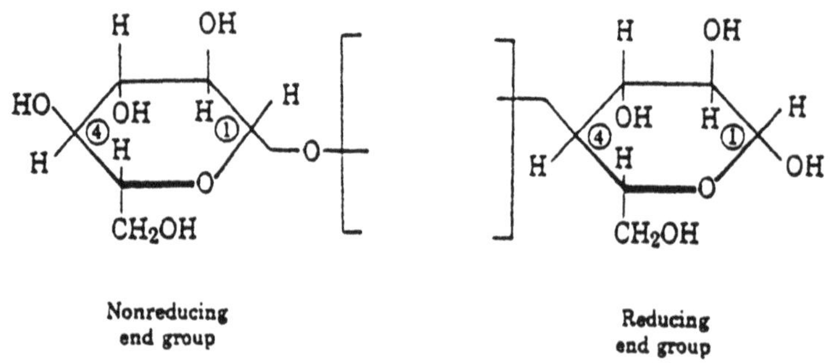

H    OH                       H    OH

HO    OH   H    H                    OH   H    H

**Nonreducing
end group**

                                    **Reducing
end group**

**FIGURE 5.4**   End groups in a cellulose molecule.

microscopic level, cellulose is found to be made up of fibrils that are about $10-500$ µm in diameter. The larger values may be due to the aggregation of several microfibrils. The average diameter of microfibrils depends on the origin of cellulose and increase with the order wood < cotton < bacterial cellulose < animal cellulose.

The degree of polymerization for cellulose can be determined by performing hydrolytic degradation of cellulose, while the chain length of cellulose can be determined by knowing the degree of polymerization.

$$\text{Chain length} = \text{DP avg. (5.15A)} \qquad (5.1)$$

where 5.15 is the length of a glucose unit.

Another important aspect of cellulose structure relating to degradation is the extent of crystallinity. It is found from X-ray diffraction studies that cellulose has at least four well-defined crystal structures. They are cellulose I for native celluloses, cellulose II for hydrated celluloses, cellulose III from a combined structure of cel-

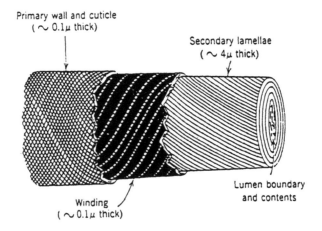

**FIGURE 5.5**   Macroscopic illustration of cellulose fiber.

lulose I and cellulose II, and cellulose IV from a combined structure of cellulose I and II (a different configuration than cellulose III). The degree of crystallinity of cellulose varies from 70% for native cellulose to 40% for hydrated celluloses. The disordered or amorphous regions have a large effect on the mechanism of cellulose degradation, which is discussed later in this chapter.

## 5.2.1 Thermal Degradation of Cellulose

Thermal degradation of cellulose under a pyrolytic environment is important because it is an important source of char generation and, therefore, has significant impact on the modeling of biomass gasifiers as well as pyrolysis units. A typical nonisothermal cellulose pyrolysis thermogram obtained at 10°C/min heating rate is given in Figure 5.6. The most acceptable overall cellulose pyrolysis scheme has been proposed by Bradbury et al. (1979) and is shown in Figure 5.7. However, since cellulose pyrolysis is a complex process, it is best to study its thermal degradation in several stages to appreciate some of the intricacies.

The pyrolysis of cellulose can be divided into three stages. The first one takes place below 250°C, where volatilization of the cellulose polymer is slow and is affected markedly by the structure of cellulose fiber. At temperatures above 250°C, the cellulose polymer begins to decompose rapidly, forming condensable volatiles along with permanent gases like carbon dioxide, carbon monoxide, and

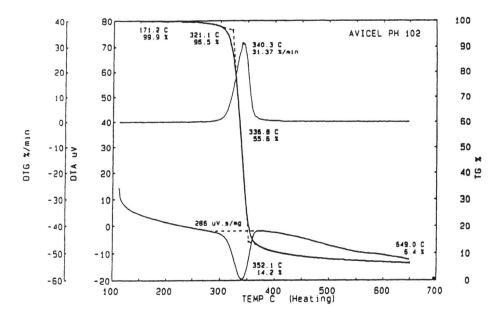

**FIGURE 5.6** A typical nonisothermal cellulose pyrolysis thermogram.

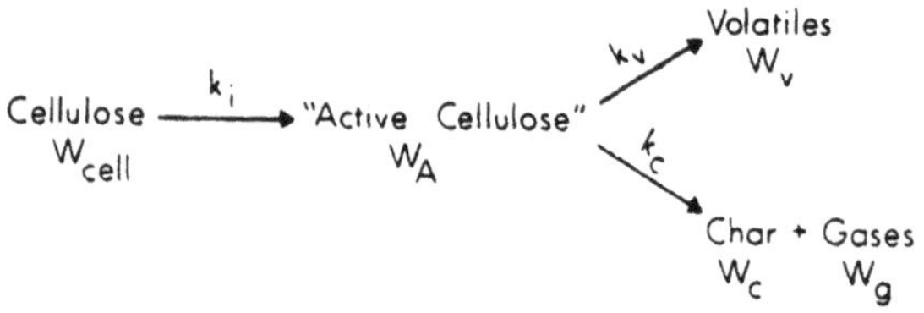

**FIGURE 5.7** Bradbury and Shafizadeh's model for cellulose pyrolysis.

water. This stage signifies the liberation of majority of the volatile products from cellulose and continues till about 500°C. Above 500°C, the volatile products undergo gas-phase pyrolysis, producing a variety of permanent gases. This stage is primarily responsible for the contribution of large hydrocarbon compounds to the final product gases due to the presence of cellulosic matter in a biomass sample. In this section, the kinetics and mechanism for the pyrolysis of cellulose has been categorized according to these temperature regimes.

## 5.2.2 Depolymerization of Cellulose at Low Temperatures

When dry cellulose particles are subjected to low-temperature pyrolysis, they undergo depolymerization process resulting in a rapid drop of *degree of polymerization* (DP). This change causes significant effects on the structure and cross linking of the cellulosic fiber and on its pyrolysis rates and product formation. The presence of trace quantities of extraneous materials like mineral matter from ash in this temperature range usually has a pronounced effect on the observed pyrolysis phenomena of cellulosic materials. However, in this section we have considered studies that were carried out on pure cellulose samples to determine the kinetic parameters for cellulose pyrolysis.

The decomposition of cellulose begins at temperatures as low as 70°C. In a low-temperature pyrolysis study of cellulose, Golova (1975) observed that the cellulose DP reduces to about 200 from 1000, and that the reduction in DP is accompanied by the formation of levoglucosan during the first $5-10\%$ weight loss of the sample. The depolymerization process was observed to take place up to 20 to 30% weight loss of the sample. The major products of cellulose pyrolysis at temperatures below 250°C are $H_2O$, CO, $CO_2$, and char. The fibers of cellulose decompose at around 100°C to give $90-95\%$ $CO_2$ and $5-10\%$ $H_2O$ on a volume basis. Shafizadeh (1975) measured the rates of formation of CO and $CO_2$ in $N_2$ and air at 170°C, and has shown that the rate of carbon bond scission greatly exceeded the rates of CO and $CO_2$ evolution in an inert atmosphere.

Since CO is the major constituent of producer gas formed due to char gasification, the understanding of its evolution in the pyrolysis stage becomes important for gasifier design. In the pyrolysis unit, this CO is part of the many combustible gases. In order to determine the process for the evolution of CO during the early stages of pyrolysis, experiments were conducted with carboxycellulose to determine whether evolution of CO was associated with the decarboxylation process of carbonyl and carbonyl group formation at temperatures below 200°C. Both of these studies concluded that there is no evidence for the formation of carbonyl and carbonyl groups in the liberation of CO, which would lead us to say that CO liberation is a complex process associated with different steps. However, the overall fraction of CO contribution due to cellulose pyrolysis is less than 10% by volume.

Studies by Patai and Halpern (1970) on different cellulose samples at later stages show that the residue remaining after 20–30% weight loss is mainly composed of oligosaccharides, suggesting that the early stages of cellulose weight loss does not involve the destruction of the carbohydrate structure of cellulose. In addition, Madorsky and co-workers (1956, 1964) found that the rates of formation of $H_2O$, CO, and $CO_2$ gradually decreased at this stage.

## Kinetic Mechanism for Depolymerization of Cellulose

Degree of polymerization for cellulose undergoes a rapid drop from 1,000 to about 200 during the early stages of pyrolysis (Golova et al., 1959). Paucault and Sauret (1958) confirmed these findings and showed that the decrease in DP follows a zero-order rate law with an activation energy of 25.8 kcal/g mole. This means that the depolymerization process follows an unzipping process after selective cleavage. The authors in this study speculate that the bond scissions occur at the boundaries between the crystalline and amorphous regions of the cellulose. This argument is also supported by Shafizadeh and Bradbury (1979), who indicate that in $N_2$ the bond scission reaction initially follows a zero-order rate law with an activation energy of 27 kcal/g mole. However, Fung (1969) reported similar findings but concluded that the reduction in DP followed a first-order rate law with an activation energy of 35.4 kcal/g mole. According to this study, the depolymerization process is the result of random scission and unzipping of the polymer does not take a lead role. The difference in the two studies could be due to temperature region and heating rate at which the studies were performed. The study carried out by Fung was conducted at relatively high heating rates and it is well known that if extra energy is provided during the depolymerization process, it could result in random scission of the bonds and mask the unzipping process.

Halpern and Patai (1969) found that at temperatures around 250°C, the DP for crystalline cellulose dropped from 1800 to 200 very rapidly and then stabilized. The overall weight loss of the cellulose sample for this drop in DP amounted to about 10 to 15%. However, similar studies with an amorphous cellulose caused only a small decrease in DP. The decomposition of cellulose at low temperature is also discussed by Chatterjee and Conrad (1966), who suggest that the cellulose breakdown at low temperatures follows a random chain cleavage followed by stepwise depolymerization mechanism. However, no detailed reasoning for their conclusion has been provided.

The primary influence of cellulose crystallinity apart from DP reduction during pyrolysis is on the extent of char formation and its relation to the location of the cleavage for the bonds at the initial stages. Since the decrease in degree of polymerization followed the development of crystallinity, one can conclude that bond rupture occurs at points of maximum strain in the polymer along the boundary where the crystalline and amorphous regions meet. This bond rupture leads to dehydration of the cellulose polymer.

Broido and Weinstein (1970) show that decrystallized cellulose forms less char than regular cellulose and that its initial rate of weight loss exceeded with crystalline cellulose. In addition, it was found that the crystalline regions of the material were the major source of levoglucosan. Because the measured crystallinity index of cellulose showed little drop even when weight loss reached 59%, Broido and Weinstein argued that unzipping of the polymer must proceed along chains through both the ordered and disordered regions of the material.

In another study, Kato and Kamorita (1968) offer an interesting insight into the effects of crystallinity on the formation of volatile compounds. Comparing the pyrolysis products of crystalline and amorphous cellulose between 200 and 300°C, the yields of furfural and 5-hydroxy-methyl furfural were observed to increase more rapidly with time during the pyrolysis of the amorphous cellulose sample, whereas at 500°C higher yields of acetaldehyde were obtained from the crystalline cellulose. These results support the idea that different reaction mechanisms take place during the pyrolysis of crystalline and noncrystalline cellulose.

The activation energy of pyrolysis for crystalline cellulose is in the neighborhood of 55–62 kcal/mole and for noncrystalline cellulose in the range of 25–30 kcal/mole. The difference is almost double, which tells us how different the rates of pyrolysis for the two celluloses would be.

### 5.2.3  Effects of Polymer Orientation/Cross-Linking on Low-Temperature Pyrolysis of Cellulose

Basch and Lewin (1973) investigated the effects of orientation of a cellulose polymer on the rate of pyrolysis and char formation. The authors found that the rate of the initial reaction, described as thermal cross linking, increased with increasing orientation according to the first-order Arrhenius rate law and that this increase in the reaction rate resulted in the increased formation of char residue. It should be noted that increased cross-linking results in increased crystallinity, which also results in increased char formation. Hence both aspects relate to similar conclusions.

A second source of char formation in the noncrystalline region is the slower bulk decomposition due to chain scission. To explain these results, Basch and Lewin (1973) argue that with low orientation the distance between chains is comparatively large and hence the rate of cross linking is low; however, with low orientation at elevated temperatures, the chain is more free to align itself to favor cross linking, thus increasing the overall extent of the cross-linking reaction. This cross linking may be due to the cleavage of a hydrogen atom from a carbon atom. The free hydrogen atom could then strip a hydrogen atom from an adjacent carbon chain, leaving two chains with highly reactive sites that could interact to form an intermolecular bond or cross link. Broido (1969), however, suggests that with increased temperature, the swelling procedure causes an increase in the distance between chains and thereby artificially reduces the rate of cross-linking reactions in the decrystallized cellulose.

From the point of view of thermal gasification of cellulosic substances, two aspects emerge from these studies. One is that the crystalline cellulose found in a daily routine would provide high yields of char that needs to be gasified during the char gasification zone of the gasifier and thus would result in the increased residence time for the sample. Second, that the activation energy of the crystalline cellulose would be much higher than the noncrystalline cellulose due to the increased cross linking. This suggests that the heat load in the pyrolysis zone for crystalline cellulosic materials will be higher. This higher activation energy, however, has its own advantage, that is, the pyrolysis product gases would have relatively less hydrocarbons because of increased thermal cracking of volatiles. The activation energy for thermal cracking is in the range of 42–49 kcal/mole. This was, however, different in the case of hemicellulose pyrolysis, where the pyrolysis activation energy was much less in comparison to the thermal cracking of the volatiles and hence resulted in increased uncracked volatiles.

### 5.2.4 Pyrolysis of Cellulose Under Moderate Temperature Conditions

At temperatures above 250°C, cellulose rapidly undergoes complete degradation by pyrolysis, forming a variety of permanent gases, condensable liquids, and char. During this temperature region char formation is the major step.

Shafizadeh postulated a global model for cellulose pyrolysis as shown in Figure 5.7. This model has been expanded (Fig. 5.8) to provide more details of cellulose pyrolysis. It can be seen that the formation of levoglucosan and anhydrosugars is an important step at low temperatures. However, at increased temperatures, the dehydration and charring reaction progresses more rapidly than reactions for transglycosylation and levoglucosan formation, resulting in increased char, CO, $CO_2$, and $H_2O$ from the cellulose substrate. This has been confirmed by Diebold's (1993) model, discussed later. Figure 5.9 shows the increase in char formation with heating rate as predicted by Diebold's model. It is appropriate to mention here that Varhegyi et al. (1995) recently provided the new cellulose kinetic model, in which he has taken away the step involving the active cellulose formation at low heating rates. Figure 5.10 gives the recent cellulose pyrolysis mechanism as proposed by Antal. This scheme has, however, been challenged by Lede et al. (1997), who present experimental evidence for the presence of some liquid phase that can be categorized as active cellulose at high heating rates.

The products of cellulose pyrolysis in this region are characterized as primary or secondary, according to whether their immediate precursor was the cellulose substrate or another product of cellulose pyrolysis.

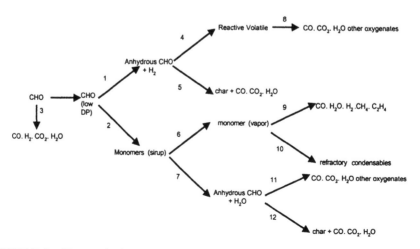

**FIGURE 5.8** Expanded scheme of cellulose pyrolysis.

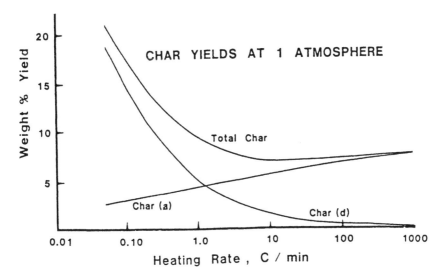

FIGURE 5.9  Char formation as a function of heating rate.

Since the products of cellulose pyrolysis and that of levoglucosan are very similar, most of the cellulose decomposition models use the formation of levoglucosan as the primary step in the cellulose decomposition. This has been supported with the findings of Paucault and Sauret (1958), who have identified levoglucosan as a major product of cellulose pyrolysis. This is not to suggest that the entire cellulose sample goes through levoglucosan formation, because some of the cellulose pyrolysis products such as phenols and cresols do not come from levoglucosan.

The two viewpoints expressed in the literature concerning the mechanism of formation of levoglucosan are as follows: The first asserts that the glycosidic bonds are ruptured homolytically and that depolymerization proceeds by a free radical mechanism. This view in part is shared by Madorsky (1958), who says that a high

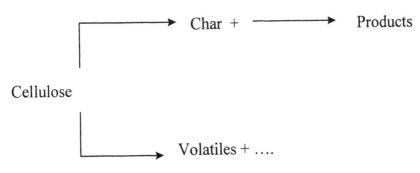

FIGURE 5.10  Antal's (1995) new mechanism for cellulose pyrolysis.

yield of monomer from a polymer chain indicates an unzipping sequence that is primarily the result of free radical reactions. The second view, proposed by Kilzer and Broido (1965), regards the transglycosylation mechanism to be heterolytic with the depolymerization process proceeding by a carbonium ion intermediate. Both points of view are based on experimental evidence. The heterolytic bond cleavage mechanism of levoglucosan results in the formation of 1,4 anhydro-D-glucopyranose, which quickly rearranges to form either 1,6-anhydro-6-D-glucopyranose (levoglucosan) or 1,6-anhydro-6-D-glucofuranose.

Bryne et al. (1966) described an extensive series of experiments on the pyrolysis of polysaccharides formed during cellulose pyrolysis. Noting that the pyrolysis products resemble those derived from acid catalyzed reactions in aqueous solutions, they propose two general pyrolysis modes. The first involves concerted displacements and the formation of an intermediate 1-2 anhydroglucose unit. This unit further decomposes to form either 5-(hydroxymethyl) furfural, 1,6-anhydro-6-D-glucofuranose, or levoglucosan. The second mode results in the formation of volatile carbonyl compounds or condensation products (char) by an irreversible carbonium ion mechanism.

Very compelling evidence supporting the free radical mechanism was given by Kislitsyn (1971), who introduced a free radical inhibitor (di-,6-naphthylphenylenediamine) into the cellulose macromolecule and observed its effects on the product distribution. Kistitsyn et al. found that the introduction of 2.25 mole% inhibitor reduced levoglucosan formation by more than 50%, and 4.5 mol% inhibitor suppressed formation almost entirely.

Another aspect of cellulose pyrolysis in this region relates to char formation. Since the char formed during the pyrolysis process will be gasified in a gasification process, it is important to know its reactivity for proper design of the reactor. Shafizadeh notes that the char formed under these conditions is pyrophoric and has extraordinary gasification rates relative to coal- and peat-derived chars.

The dependence of char yield on the temperature of pyrolysis was also shown by the work of Broido and Nelson (1975), who pretreated cellulose samples at temperatures ranging between 230 and 275°C prior to pyrolysis at 350°C. Depending upon the temperature and duration of the pretreatment, the char yield varied from 10 to 27% by weight. They showed that this behavior could easily be explained by a competitive degradation mechanism involving depolymerization and dehydration reactions. It was found that the depolymerization reaction formed little or no char, whereas the dehydration reaction formed 36% by weight char. Furthermore,

Broido and Nelson showed the following ratio of reaction rates for these two reactions, $k_2/k_2 = (2.05 \times 10^4)\exp(-8,900/T)$.

## 5.2.5  Effects of Rapid Heating on Cellulose Pyrolysis

At high heating rates due to shift in reaction temperature, as explained by the Arrhenius equation, cellulose pyrolysis takes place at temperatures over 400°C. Berkowitz-Mattuck and Noguchi (1963) noted the decrease in char formation from cellulose with increasing heating rate, while Martin (1965) observed negligible char formation at very high heating rates, and concluded that at least one pathway exists for cellulose pyrolysis that does not include char formation.

Berkowitz-Mattuck and Noguchi (1963) note the presence of $CO$, $CO_2$, $CH_4$, levoglucosan, and at least 12 polar organic compounds in the volatile products with radiant flux densities of 20 to 100 W/cm$^3$. Martin (1965) found $CO$, $CO_2$, and $H_2O$ to be the earliest volatile products. Increasing radiant flux densities effected increasing yields of saturated and unsaturated aldehydes, ethylene, and hydrogen while decreasing yields of $H_2O$ and $CO_2$. Martin asserts that levoglucosan is the only product of the higher-temperature pyrolysis pathway, concluding that the presence of $H_2$, $CH_4$, $C_2H_6$, $C_3H_8$, and other hydrocarbons in the gaseous products is a result of secondary, heterogeneous pyrolysis reactions between the levoglucosan vapors and the hot char.

In his studies of the effects of the intensity of radiant energy on cellulose pyrolysis, Lincoln (1980) notes an increase in the mass ratio of $CO$ to $CO_2$ formation from 0.33 at 6 W/cm$^3$ to 9.2 at 12,000 W/cm$^3$ (peak). Lincoln attributes this increase to the fact that flash heating indiscriminately breaks the cellulose polymer into small pieces. Baker 91975) confirms this finding and speculates that the favorable influence of heating rate on $CO$ formation is due to increasing fracture of the glycosidic ring. The shock tube studies of Eventova et al. (1974) provide further evidence for the enhanced formation of $CO$ and $H_2$ at higher temperatures.

Milne and Soltes (1981) have reported the results of some very exciting research involving the flash pyrolysis of a great variety of biomass materials in hot steam (to 900°C or more). This work parallels that of Schulten et al. (1981), but involves studies of a wide range of experiment conditions intended to elucidate the high-temperature pyrolysis pathways. Interestingly, Milne and Soltes note that all of the higher-mass species seem to evolve nearly simultaneously as though they arose from a common intermediate, which in turn corroborates Martin's hypothesis, which states that levoglucosan is the only the primary product and the rest are due to secondary pyrolysis.

## 5.2.6 Determination of Kinetic Parameters for Cellulose

In 1969, Halpern and Patai discussed the futility of attempts to employ a simple rate law in modeling the complex mechanism of cellulose pyrolysis. Their results led them to conclude that if a simple rate law were presumed then the evaluation of activation energies and other kinetic constants would be influenced more by the mathematical method of analysis than the actual chemistry. These caveats have been underscored by the research of Cardwell and Luner (1976) who observed a continuous variation in the values of the apparent activation energy and order of cellulose pyrolysis by using sophisticated multiple-heating-rate methods of kinetic analysis.

The determination of kinetic parameters over a narrow temperature range introduces a large uncertainty into the values of the calculated activation energy and the other rate constants. A clear example of these difficulties can be found in the analysis of data obtained from a set of experiments performed by Lipska and Parker (1966). They studied the temperature range 250–300°C and concluded that cellulose undergoes pyrolysis in three stages. At first, a rapid initial decomposition takes place, which is followed by a zero-order volatilization reaction with activation energy $E$ = 42 kcal/g mole. This is followed by a first-order char forming reaction with $E$ = 42 kcal/g mole. The pre-exponential term for this temperature region lies in the range of $1 \times 10^9$. Upon re-examining the data of Lipska and Parker in the light of a chain reaction mechanism, Chatterjee (1968) found activation energies of 49 kcal/g mole for the initiation reaction and 42 kcal/g mole for the propagation reaction. Antal et al. (1980) studied cellulose pyrolysis in the temperature range of 320–460°C using TG. Since the objective of the study was to develop an engineering model, a simple rate law was employed that did not account for the influence of temperature on char formation. Kinetic parameters were obtained that resulted in a good fit of calculated weight-loss curves with experimental data. Unfortunately, when these data were used for temperatures associated with the higher heating-rate weight-loss curves, the curves did not show a good fit. The authors concluded that this method was resulting in low activation energy. Thus this result illustrates one difficulty that can be encountered in attempting to extend the temperature range over which pyrolysis-rate data are obtained.

Table 5.3 summarizes the values of $E$ for cellulose pyrolysis obtained by a variety of workers using small (<1 g) samples. The

**TABLE 5.3**

Summary of the Values of *E* and *A* for Cellulose Pyrolysis

| Activation energy (kcal/mole) | log *A* (*A* in 1/min) | Temperature (°C) | Reference |
| --- | --- | --- | --- |
| 15.0 | 6.0 | 25–700 | Murty (1972) |
| 17.0 | 5.4 | 330–400 | Barooah and Long (1976) |
| 26.0 | 11.5 | 110–220 | Stamm (1956) |
| 32.1 | 12.9 | 145–265 | Kujirai and Akahira (1925) |
| 53.5 | 18.8 | — | Akita and Kase (1967) |
| 56.0 | 19.0 | 308–360 | Tang and Eickner (1968) |

wide range in values for *E* exhibited in the table may reflect (1) the complex pyrolysis mechanism, which cannot be adequately simulated by a simple rate law; (2) the possible influence of fine structure, ash, or other impurities on the pyrolysis rate; and (3) the narrow temperature range examined by many of the experiments.

In agreement with the literature that simple rate laws may not be able to explain the kinetic behavior of cellulose pyrolysis, Diebold (1993) proposed a seven-step kinetic model for cellulose pyrolysis (Fig. 5.11). This model incorporates the major steps involved in cellulose pyrolysis and is able to predict the degradation pattern over a wide range of heating rates. The limitation of this model lies in the fact that there are too many kinetic parameters to account for cellulose degradation. In an attempt to answer the question of changing kinetic parameters due to heating rate, Chornet and Roy (1980) proposed the compensation effect, where it was shown that

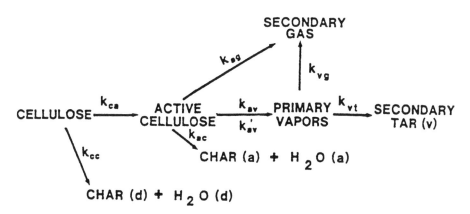

**FIGURE 5.11** Diebold's seven-step global model for cellulose pyrolysis.

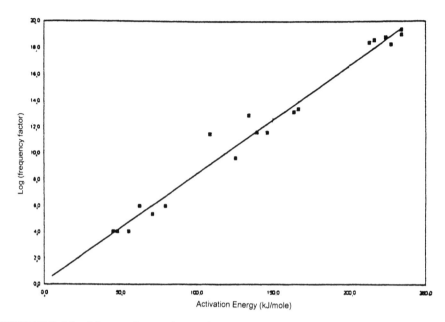

**FIGURE 5.12**  Linear dependency between log *A* and *E*.

the variation in log *A* and *E* with temperature followed a linear dependency (Figure 5.12). This effect was also noted by Reed (1981).

Later, Gaur and Reed (1994) found that by introducing the temperature dependency term to the pre-exponential factor for the nonisothermal kinetic measurements, one can have reasonable predictions on cellulose decomposition over a wide range of heating rates. This term was introduced by incorporating the concepts of isothermal kinetic measurements proposed by Laidler (1969). The application of this equation has also been validated by incorporating Diebold's seven-step kinetic model (1993).

In the end, it can be concluded that the kinetic equation for cellulose pyrolysis at low temperatures can be written by using the Arrhenius form of equation. There are strong indications that cellulose pyrolysis in this temperature regime follows a zero-order degradation pattern, at least in the initial stages of degradation (10–15% weight loss). In the later stages, in most cases, the cellulose conversion follows a first-order kinetic equation. The activation energy for depolymerization of noncrystalline cellulose has been reported over the range of 25–35 kcal/mole and that of crystalline cellulose in the range of 55–65 kcal/mole. The pre-exponential term is reported in the range of $1 \times 10^{9}$ to $1 \times 10^{12}$ in the literature. The difference in the activation energy could be due to many factors, such as the crystallinity, orientation, and degree of polymerization. In addition, experimental parameters such as heating rate and particle size also lead to difference in the determination of kinetic parameters. Based on various inputs, in our judgment is that it is best to take the activation energy and frequency factor values of the higher end in order to model the thermal conversion unit. The reasons for this recommendation are:

1. Most samples in the real world would be crystalline.
2. Samples would be exposed at relatively high heating rates, which causes the apparent activation energy to increase.

The overall cellulose pyrolysis can be considered as first-order reaction and can be described by using the Arrhenius type of equation. The kinetic equation for this reaction can be written as $k = 1 \times 10^{10} e^{-50000/RT}$.

*Kinetics and Mechanism of Cellulose Pyrolysis at Rapid Heating Rates*

High-temperature pyrolysis kinetics reflects the rates of the lower-temperature pyrolysis. Consequently, rate laws describing high-temperature pyrolysis phenomena must necessarily incorporate the competitive reaction scheme discussed earlier. In addition, the elucidation of a comprehensive high-temperature rate law also requires a determination of the rate law associated with the postulated high-temperature fission pathway. Unfortunately, no studies with regards to the effect of temperature on this reaction are available. Moreover, the mathematical complexity of three competitive pathways, combined with the recognition that the primary products of the high-temperature pathway (CO and $H_2$) are also products of the low-temperature pathway (CO) and the moderate-temperature pathway (if levoglucosan undergoes further pyrolysis in the gas phase), suggests some of the difficulties that must be overcome before a rate law for the high-temperature pathway can be determined.

The works referred to here indicate that the dominant pathway for cellulose pyrolysis at higher temperatures is the same as that at moderate temperatures, and leads to levoglucosan formation. Consequently, the questions raised earlier also pertain to high-temperature pyrolysis in terms of kinetic parameters and their determination. Based on the works of Diebold (1993), Gaur and Reed (1994), and Antal (1995), we are in a position to say that one kinetic parameter can be used to explain the global degradation of cellulose pyrolysis for moderate- and high-temperature ranges. The values represented by these works suggest that the activation energy is in the range of 55–65 kcal/mole and the pre-exponential term is around $10^{15}–10^{18}$ sec$^{-1}$. The overall kinetic equation for cellulose pyrolysis for moderate- and high-temperature regimes can be written as:

$$K = 1 \times 10^{16} e^{-62000/RT}$$

## 5.2.7 Effects of Reaction Parameters

Earlier in this chapter, much emphasis was given to the effects of temperature on the products and mechanisms of cellulose pyrolysis. For this reason, it is of interest to identify the effects of various other reaction parameters, such as heating rate, pressure, particle size, ambient gas environment, and additives (catalysts) on the products, pathways, and rates of cellulose pyrolysis.

Since cellulose is highly reactive, it is difficult to heat it rapidly enough to cause isothermal pyrolysis at temperatures in excess of about 600°C. Consequently, above 500°C the effect of heating rate can be of greater interest to an engineer than the effects of temperature. In general, a higher heating rate favors the higher-temperature transglycosylation reactions over the lower temperature charring reactions.

The effects of pressure on biomass pyrolysis have been described in a series of papers by Mok and Antal (1981). In general, increasing the pressure dramatically reduces the yield of volatile materials and enhances the formation of char whereas decreasing pressure favors the volatile formation. An increase in ambient pressure from 0.1 to 2.5 MPa increases the char yield from 12% to 20% by weight. Among gaseous products, the formation of $CO_2$ and $H_2$ is favored by increasing pressure, while the yields of CO, $CH_4$, $C_2H_2$, and other light hydrocarbons are reduced. Under high pressure (0.5 MPa), an increased residence time reduced the yield of CO and hydrocarbons, and enhanced the formation of $CO_2$ in the product stream.

A strong correlation exists between the heat of the reaction and the char yield. An increase in char yield is associated with an increase in reaction exothermicity. Consequently, increasing pressure is also associated with increasing exothermicity of the pyrolysis reactions. However, the relationship is complicated by the heats of reach of the gas-phase pyrolysis reactions, which are also influenced by pressure. Since increasing pressure favors exothermicity in the gas-phase reactions, a DSC may detect equal exothermicity under high pressure and high flow rate conditions, or moderate pressure and low flow rate conditions.

Using both the techniques of thermogravimetry and DSC, Arseneau (1971) was able to observe the appearance of a high-temperature exotherm as cellulose samples of increasing thickness were studied. This exotherm was attributed to the formation of secondary char resulting from the decomposition of volatile matter levoglucosan that was unable to rapidly escape from the thicker cellulose samples. Thus, the pyrolysis of larger particles of cellulose should evolve more char and heat at the expense of volatiles formation. Pyle and Zorar (1984) presented a model for cellulose pyrolysis as a function of internal heating of the sample. They have derived a nondimensional number, called the Pyrolysis number (along the lines of the Biot number, which represents the ratio of external heat transfer to internal heat transfer within the sample. It is found that if the Pyrolysis number is less than 1, then it results in more char yield.

Biomass pyrolysis can be accomplished practically in a variety of gaseous environments. Steam is thought to be the most practical alternative; however, hydrogen, methane, and recycled gas have also been studied.

The effects of steam on cellulose pyrolysis have been studied, but are not yet well understood. The early work of Wiegerink (1940) and Waller et al. (1948) indicated that the thermal breakdown of cellulose is accelerated by the presence of water.

The low-temperature results of Stamm (1956) evidenced much more rapid degradation of cellulose under steaming conditions, reportedly lowering the apparent activation energy.

The dramatic effects of trace amounts of ash on cellulose pyrolysis were noted earlier in the beginning of the cellulose section. Milne (1981) describes the dramatic formation of 50% char from rapidly heated cellulose impregnated with 10% (by weight) K+ ions derived from $K_2CO_3$ or KOH. The char formation was accompanied by the prompt evolution of low-molecular-weight gases. Thus, the alkaline catalyzed, rapid heating of cellulosic materials may be an attractive source of char.

Low heating rates, increased pressure, and the presence of ash or additives all enhance the formation of char at the expense of volatile production. The relative ease of transforming biomass into charcoal justifies an exploration of the possibilities for using gasification as the mode for energy transformation rather than pyrolysis to form liquid fuels.

## 5.3   LIGNIN

Lignin forms about 30% by weight of a biomass sample and is responsible for providing the structural strength to the sample. It is usually concentrated between the space of the cells where it is de-

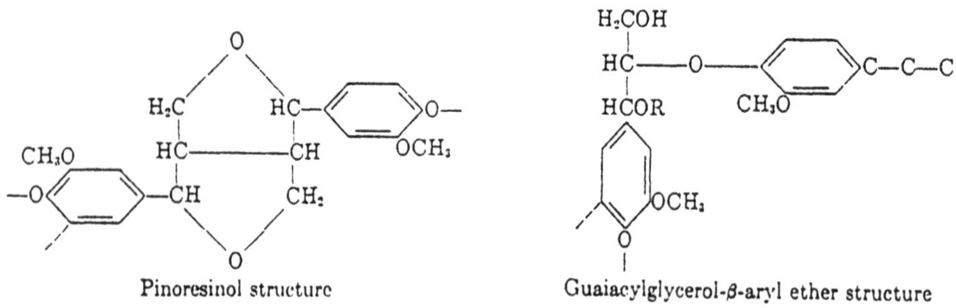

**FIGURE 5.13** Structure of a lignin molecule.

posited during the lignification process. The *lignification* of a cell denotes the end of living cell functions and is, therefore, considered an irreversible end product. Plant lignins can be divided into three classes: gynosperm (softwood), angiosperm (hardwood), and grass lignins. The chemical structure of lignin shows that it is primarily aromatic in nature. This aromaticity of lignin was first demonstrated by Lange in 1944 when he showed that spruce wood lignin exhibits a characteristic ultraviolet absorption spectrum similar to the guaiacylpropane model compound, which is aromatic in nature. Later, it was demonstrated that the majority of lignin is composed of phenylpropane polymer (Figure 5.13).

In a lignin macromolecule, the monomeric phenylpropane units are linked together both by ether and by carbon-to-carbon linkages. These carbon-to-carbon linkages provide high resistance towards chemical degradation of a lignin molecule. Figure 5.14 gives the postulated linkages in a lignin molecule. During thermal degradation, or any other kind of degradation process, lignin does not de-

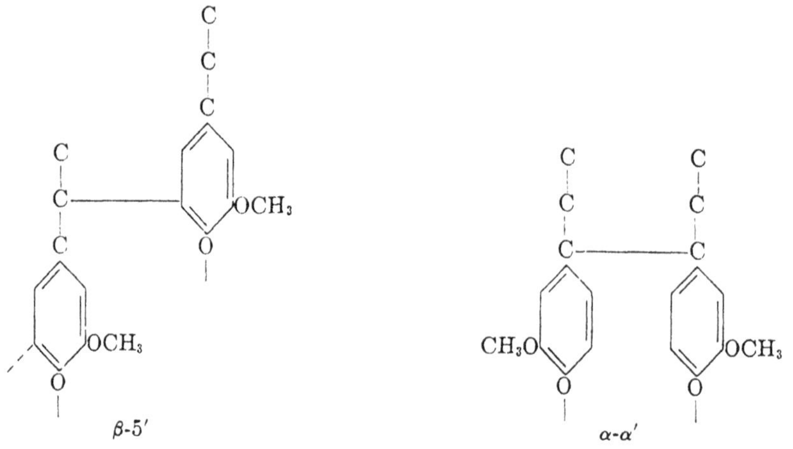

**FIGURE 5.14** Postulated linkages in a lignin molecule.

grade to monomers mainly because part of the phenylpropane units are joined by carbon-to-carbon linkages. To further classify the polymeric representation of lignins, it can be said that a majority of softwoods (gymnosperms) contain guaiacylpropane units while a majority of hardwoods (angiosperms) contain guaiacylpropane and syringlpropane units.

## 5.3.1 Pyrolysis of Lignin

Thermal degradation of lignin under pyrolytic conditions show that the volatile products from lignin are liberated over a wide temperature range, beginning from approximately 200°C and lasting up to 600°C. At temperatures below 200°C, some thermal softening of lignin has been observed but there is no significant weight loss. However, this statement should be used with discretion because studies have shown that lignins isolated from softwoods are more stable than those from hardwoods, reflecting the higher thermal stability. Figure 5.15 shows the pyrolysis thermogram for a lignin sample.

Decomposition of lignin is important to both pyrolysis and gasification kinetics because it is the major contributor to the char formation and liberates a large fraction of volatiles, which constitute tar. Both of these aspects severely affect the product gas composition. One aspect that the reader must keep in mind is that char conversion becomes important in biomass gasification not because it is the rate limiting step but because it controls the gasifier sizing.

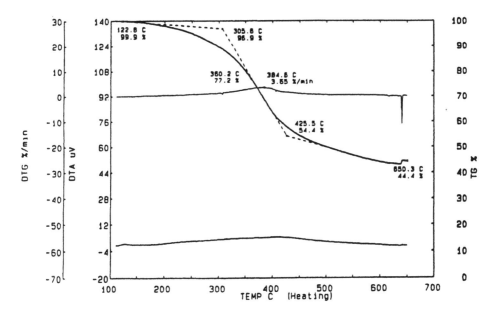

**FIGURE 5.15** Pyrolysis thermogram of Kraft lignin.

This is in spite of the fact that biomass has only 20% fixed carbon and 80% volatile matter on an ash-free basis.

At somewhat higher temperatures (240–380°C), lignin undergoes a devolatilization process and a variety of gases along with condensable liquids are liberated. This transformation results in char formation. The major difference in lignin pyrolysis and cellulose pyrolysis lies in the fact the cellulose decomposition has a defined reaction pathway where levoglucosan is formed, while in the case of lignin such a path has not been identified. However, the products formed during lignin pyrolysis are characteristic to the feed material. For example, upon careful examination one can differentiate the char formed by the pyrolysis of lignin obtained from maple or oak wood. This suggests that by knowing the feed properties of the biomass one can estimate the char reactivity to certain extent because lignin is the major source of char. This aspect of lignin pyrolysis can be used to advantage during the design step of any thermal conversion unit.

## 5.3.2   Pyrolysis of Lignin at Low Heating Rates

At low heating rates, the char yield from lignin can be as high as 50% by weight. This char is insoluble in organic solvents, acids, and alkalis, and not susceptible to hydrocracking or hydrotreating, even in the presence of catalysts.

Water, methanol, acetic acid, acetone, and acetaldehyde are major components of the aqueous distillate and represent about 23% by weight of lignin substrate. Tar yields under these conditions is of the order of 15–20%. Formation of permanent gases include $CO_2$, CO, $CH_4$, $H_2$, and some hydrocarbons.

Since lignin pyrolysis does not have discrete product distribution and its pyrolysis mechanism is a function of feed characteristics, most of the lignin pyrolysis understanding is based on model studies.

Thermal decomposition of aspen lignin shows that its decomposition is influenced by the reaction temperature as well as heating rate, indicating that several different free radical species are involved in the reaction process.

Since lignin structure is highly aromatic in nature and represents those of phenolic compounds, Domburgs et al. (1974) conducted studies with dehydrodiisoeugenol, a phenolic compound that liberates phenols, cresols, and related compounds upon thermal degradation. The kinetic parameters were measured using thermal analysis techniques. It was found that these model compounds degraded in the temperature range of 370–400°C. At low heating rates (10°C/min), it was found that the dehydrodiisoeugenol degraded with homolytic cleavage forming benzoyl and phenoxy radicals. The activation energy using the Arrhenius rate law was found to be in the range of 16–35 kcal/g-mole. In another work, the kinetics of lignin pyrolysis was attempted by undertaking this study with 20 selected model parameters (Klien and Virk, 1981). The study included the entire temperature range of 250–550°C. Based on the studies conducted on these model compounds, it was found that the decomposition of lignin can be successfully explained by using the Arrhenius type of rate law involving first-order reaction rate kinetics. The value of activation energy was determined to be 45 kcal/g-mole and that of the pre-exponential term to be $1 \times 10^{11}$. The major reaction products were phenol and styrene. The styrene product of the primary pathway undergoes secondary degradation reactions using free radical mechanism and forms various hydrocarbons, including ethylbenzene and toluene.

Friedman's (1964) analysis for global kinetic mechanism shows that there are series of competitive reactions that take place during lignin pyrolysis. Of these reactions, the ones that have low energies of activation contribute to char formation and the ones that have high activational energy constitute the formation of monomers of volatile gas products. Figure 5.16 shows one of the Friedman curves. From this figure, it is seen that the initial decomposition of lignin at low heating rates has a high activation process. However, once the reaction initiates there is a rapid decrease in the activation energy, which leads to the liberation of permanent gases and char formation. The increase in activation energy at a later stage depicts the presence of competitive reactions leading to monomer formation.

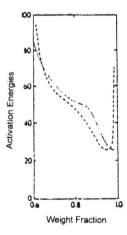

**FIGURE 5.16**  A Friedman curve.

Some researchers have also treated lignin pyrolysis as a single reaction and have reported apparent activation energies and other kinetic parameters. While some other works have calculated sets of rate parameters within specified temperature ranges with a single pathway as a dominant one. The activation energy for these studies lies in the range of $35-45$ kcal/gmol. The findings of Fenner and Lephardt (1981) suggest that the evolution profiles of all major volatile components could be explained by a mechanism with three to five degradation pathways.

### 5.3.3 Pyrolysis of Lignin Under High Heating Rates

As with other biomass components, the increase in heating rate decreases the char formation and enhances the formation of condensable products and/or gases, depending on reactor conditions. Furthermore, high temperatures favor the formation of unsaturated hydrocarbons. Using microwave radiation as a rapid heating source, Chan and Krieger (1981) observed the following yields: 33.5% char, 19.6% volatile hydrocarbons (mostly tar forming), and 36.3% permanent gases. This suggests that under a fluidized bed reactor, more tar-forming volatiles will be formed. But since the temperature of lignin decomposition will be high as opposed to one in a fixed bed reactor, these volatiles will also be subjected to elevated temperatures for thermal cracking ($E = 45-55$ kcal/mole). Hence, there is an optimization effort during the fluidized bed gasifier design process in order to minimize tar formation.

In order to test the secondary char formation during fast pyrolysis of lignin, Hopkins and co-workers (1984) conducted experiments at 400°C/sec heating rate. During this study, they found that significant amounts of condensed carbon soot was formed during this reaction process. The reactivity of this char material was very less in comparison to the primary char obtained during lignin pyrolysis.

One aspect of lignin pyrolysis found to be in common with the pyrolysis of other components of biomass is the fact that a certain fraction of CO and $CH_4$ are formed no matter what the changes are with respect to temperature or pressure. This suggests that certain carbon atoms are designated for the formation of these products. The quantification of these products by each biomass fraction becomes important in the gasification process because both of these components are the major energy carrier gases. It is estimated that the percentage of CO formation in lignin pyrolysis is in the range of $2-6\%$. No clear estimates on $CH_4$ percentage range is available except that its formation is also favored by the increase in temperature and heating rate.

## 5.3.4  Summary of Lignin Pyrolysis

At temperatures below 200°C, no major evolution of product gases is evidenced in the case of lignin pyrolysis; however, sufficient evidence of thermal softening of lignin fibers is observed, which may be associated with some weight loss (1–2%). Lignin pyrolysis commences at around 250°C and continues over a broad temperature range up to 600°C. There is no evidence that lignin thermal degradation takes place through a set reaction pathway, as is the case with cellulose, where formation of levoglucosan is almost inevitable. In contrast, lignin pyrolysis is more dependent on the initial feed material. In fact, it has been found that by knowing the final reaction products the source of lignin feed material can be identified with relatively high success.

Since lignin is the major source of char contributor during biomass pyrolysis, it has a significant impact on the gasifier and combustor design. In addition, the formation of tar is directly dependent on thermal conditions in addition to the source material. These two factors put together make the kinetics of lignin pyrolysis one of the important features during the design of any biomass conversion process.

It is found that lignin undergoes thermal degradation through several competitive reactions. The lumped kinetic parameters for lignin can be written as follows using the Arrhenius form of the first-order rate law:

$$K = 1 \times 10^{11} e^{-45000/RT}$$

Another important aspect to note with regards to lignin pyrolysis is that, at very high heating rates, a reaction pathway exists for lignin thermolysis that does not include char formation. During this process, char formation was observed to occur by the condensation of vapor-phase species apparently left in a supersaturated state after the rapid heating of the lignin substrate. This carbon soot has very low reactivity in comparison to primary char formed during low heating rates.

The effect of pressure has been shown to be favorable on char formation. Enhanced char formation is accompanied by enhanced exothermicity of the solid-phase pyrolysis reactions. This behavior resembles the response of the individual biomass components to increasing pressure. In addition, studies by Roy and Chornet (1980) have shown that vacuum pyrolysis increases the yields of tars from many lignocellulosic materials. Tar yields in excess of 50% by weight were obtained from poplar wood, with a char yield of 15%.

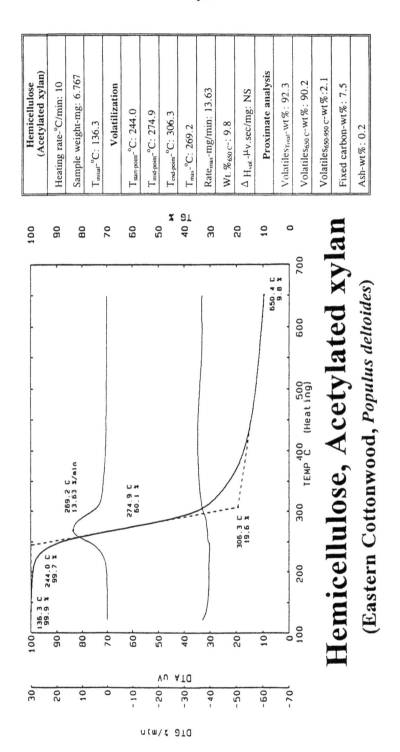

| Hemicellulose (Acetylated xylan) | |
|---|---|
| Heating rate-°C/min: 10 | |
| Sample weight-mg: 6.767 | |
| $T_{initial}$-°C: 136.3 | |
| **Volatilization** | |
| $T_{start-point}$-°C: 244.0 | |
| $T_{mid-point}$-°C: 274.9 | |
| $T_{end-point}$-°C: 306.3 | |
| $T_{max}$-°C: 269.2 | |
| $Rate_{max}$-mg/min: 13.63 | |
| Wt.%$_{650 C}$: 9.8 | |
| $\Delta H_{vol}$-$\mu$v.sec/mg: NS | |
| **Proximate analysis** | |
| Volatiles$_{Total}$-wt%: 92.3 | |
| Volatiles$_{650 C}$-wt%: 90.2 | |
| Volatiles$_{650-950 C}$-wt%: 2.1 | |
| Fixed carbon-wt%: 7.5 | |
| Ash-wt%: 0.2 | |

# Hemicellulose, Acetylated xylan

## (Eastern Cottonwood, *Populus deltoides*)

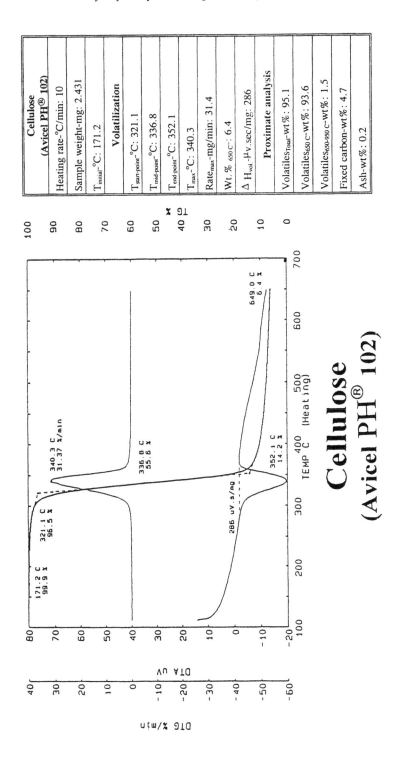

| Cellulose (Avicel PH® 102) | |
|---|---|
| Heating rate-°C/min: 10 | |
| Sample weight-mg: 2.431 | |
| $T_{initial}$-°C: 171.2 | |
| **Volatilization** | |
| $T_{start-point}$-°C: 321.1 | |
| $T_{mid-point}$-°C: 336.8 | |
| $T_{end-point}$-°C: 352.1 | |
| $T_{max}$-°C: 340.3 | |
| $Rate_{max}$-mg/min: 31.4 | |
| Wt.% $_{650°C}$: 6.4 | |
| $\Delta H_{vol}$-$\mu$v.sec/mg: 286 | |
| **Proximate analysis** | |
| Volatiles$_{Total}$-wt%: 95.1 | |
| Volatiles$_{650°C}$-wt%: 93.6 | |
| Volatiles$_{650-950°C}$-wt%: 1.5 | |
| Fixed carbon-wt%: 4.7 | |
| Ash-wt%: 0.2 | |

## Cellulose
## (Avicel PH® 102)

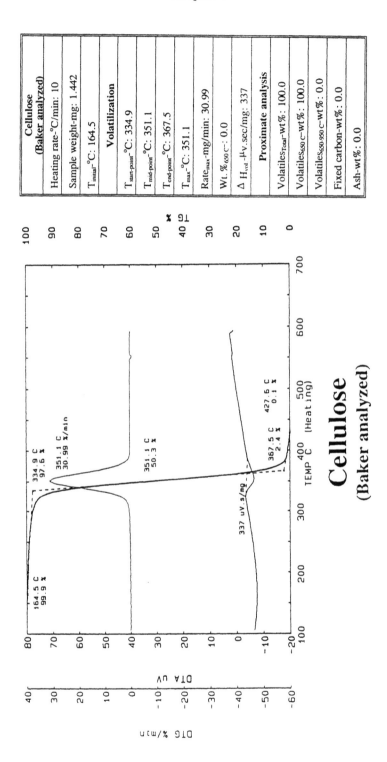

| Cellulose (Baker analyzed) | |
| --- | --- |
| Heating rate-°C/min: | 10 |
| Sample weight-mg: | 1.442 |
| $T_{initial}$-°C: | 164.5 |
| **Volatilization** | |
| $T_{start-point}$-°C: | 334.9 |
| $T_{mid-point}$-°C: | 351.1 |
| $T_{end-point}$-°C: | 367.5 |
| $T_{max}$-°C: | 351.1 |
| $Rate_{max}$-mg/min: | 30.99 |
| Wt.%$_{650C}$: | 0.0 |
| $\Delta H_{vol}$-$\mu$v.sec/mg: | 337 |
| **Proximate analysis** | |
| Volatiles$_{Total}$-wt%: | 100.0 |
| Volatiles$_{650C}$-wt%: | 100.0 |
| Volatiles$_{650-950C}$-wt%: | 0.0 |
| Fixed carbon-wt%: | 0.0 |
| Ash-wt%: | 0.0 |

## Cellulose
### (Baker analyzed)

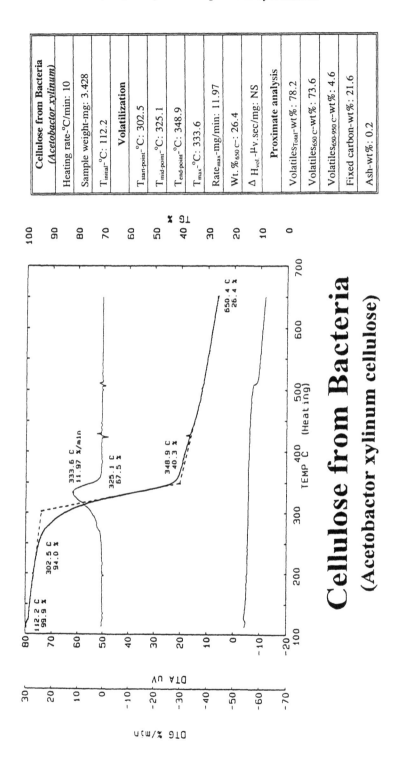

| Cellulose from Bacteria (*Acetobacter xylinum*) | |
| --- | --- |
| Heating rate–°C/min: | 10 |
| Sample weight–mg: | 3.428 |
| $T_{initial}$–°C: | 112.2 |
| **Volatilization** | |
| $T_{start-point}$–°C: | 302.5 |
| $T_{mid-point}$–°C: | 325.1 |
| $T_{end-point}$–°C: | 348.9 |
| $T_{max}$–°C: | 333.6 |
| $Rate_{max}$–mg/min: | 11.97 |
| Wt. %$_{650}$ $_C$: | 26.4 |
| $\Delta H_{vol}$–$\mu$v.sec/mg: | NS |
| **Proximate analysis** | |
| Volatiles$_{Total}$–wt%: | 78.2 |
| Volatiles$_{650}$ $_C$–wt%: | 73.6 |
| Volatiles$_{650-950}$ $_C$–wt%: | 4.6 |
| Fixed carbon–wt%: | 21.6 |
| Ash–wt%: | 0.2 |

**Cellulose from Bacteria**
**(Acetobactor xylinum cellulose)**

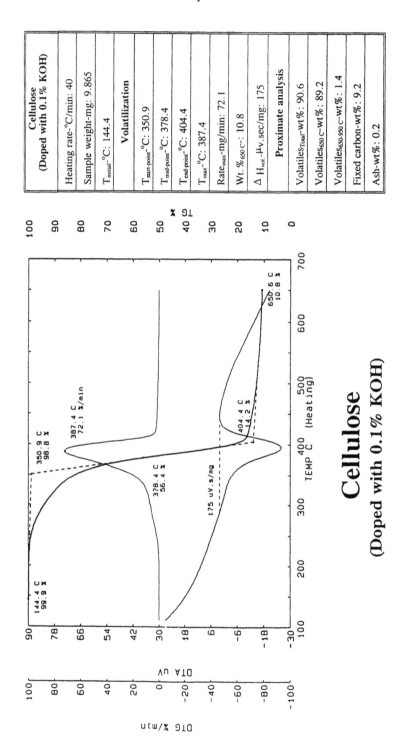

**Cellulose**
**(Doped with 0.1% KOH)**

| Cellulose (Doped with 0.1% KOH) | |
|---|---|
| Heating rate-°C/min: | 40 |
| Sample weight-mg: | 9.865 |
| $T_{initial}$-°C: | 144.4 |
| **Volatilization** | |
| $T_{start\text{-}point}$-°C: | 350.9 |
| $T_{mid\text{-}point}$-°C: | 378.4 |
| $T_{end\text{-}point}$-°C: | 404.4 |
| $T_{max}$-°C: | 387.4 |
| $Rate_{max}$-mg/min: | 72.1 |
| Wt. $\%_{650C}$: | 10.8 |
| $\Delta H_{vol}$-$\mu$v.sec/mg: | 175 |
| **Proximate analysis** | |
| Volatiles$_{Tgoal}$-wt%: | 90.6 |
| Volatiles$_{650C}$-wt%: | 89.2 |
| Volatiles$_{650\text{-}950C}$-wt%: | 1.4 |
| Fixed carbon-wt%: | 9.2 |
| Ash-wt%: | 0.2 |

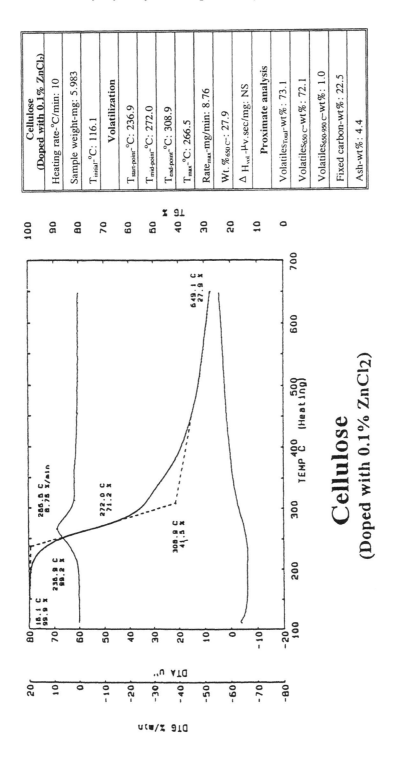

| Cellulose (Doped with 0.1% $ZnCl_2$) | |
|---|---|
| Heating rate-°C/min: | 10 |
| Sample weight-mg: | 5.983 |
| $T_{initial}$-°C: | 116.1 |
| **Volatilization** | |
| $T_{start-point}$-°C: | 236.9 |
| $T_{mid-point}$-°C: | 272.0 |
| $T_{end-point}$-°C: | 308.9 |
| $T_{max}$-°C: | 266.5 |
| $Rate_{max}$-mg/min: | 8.76 |
| Wt.%$_{650\,C}$: | 27.9 |
| $\Delta H_{vol}$.-$\mu$v.sec/mg: | NS |
| **Proximate analysis** | |
| Volatiles$_{T\,total}$-wt%: | 73.1 |
| Volatiles$_{650\,C}$-wt%: | 72.1 |
| Volatiles$_{650-950\,C}$-wt%: | 1.0 |
| Fixed carbon-wt%: | 22.5 |
| Ash-wt%: | 4.4 |

## Cellulose
(Doped with 0.1% $ZnCl_2$)

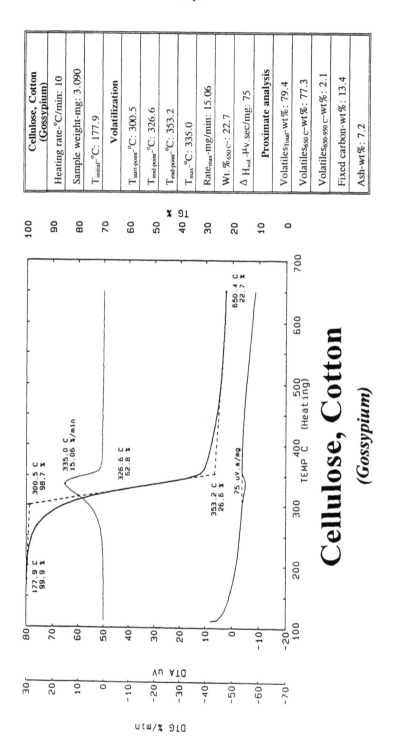

| Cellulose, Cotton (Gossypium) | |
|---|---|
| Heating rate-°C/min: 10 | |
| Sample weight-mg: 3.090 | |
| $T_{initial}$-°C: 177.9 | |
| **Volatilization** | |
| $T_{start-point}$-°C: 300.5 | |
| $T_{mid-point}$-°C: 326.6 | |
| $T_{end-point}$-°C: 353.2 | |
| $T_{max}$-°C: 335.0 | |
| $Rate_{max}$-mg/min: 15.06 | |
| Wt. $\%_{650C}$: 22.7 | |
| $\Delta H_{vol}$-$\mu$v.sec/mg: 75 | |
| **Proximate analysis** | |
| Volatiles$_{Ttotal}$-wt%: 79.4 | |
| Volatiles$_{650C}$-wt%: 77.3 | |
| Volatiles$_{650-950C}$-wt%: 2.1 | |
| Fixed carbon-wt%: 13.4 | |
| Ash-wt%: 7.2 | |

# Cellulose, Cotton
*(Gossypium)*

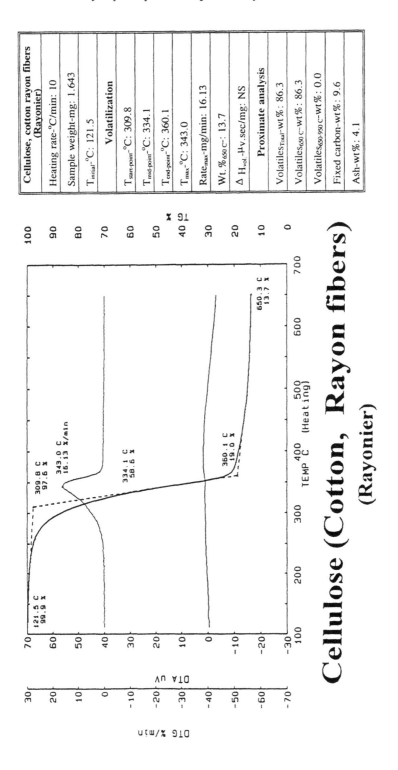

| Cellulose, cotton rayon fibers (Rayonier) |
|---|
| Heating rate-°C/min: 10 |
| Sample weight-mg: 1.643 |
| $T_{initial}$-°C: 121.5 |
| **Volatilization** |
| $T_{start-point}$-°C: 309.8 |
| $T_{mid-point}$-°C: 334.1 |
| $T_{end-point}$-°C: 360.1 |
| $T_{max}$-°C: 343.0 |
| $Rate_{max}$-mg/min: 16.13 |
| Wt.%$_{650}$ C: 13.7 |
| $\Delta H_{vol}$-$\mu$v.sec/mg: NS |
| **Proximate analysis** |
| Volatiles$_{Total}$-wt%: 86.3 |
| Volatiles$_{650\,C}$-wt%: 86.3 |
| Volatiles$_{650-950\,C}$-wt%: 0.0 |
| Fixed carbon-wt%: 9.6 |
| Ash-wt%: 4.1 |

## Cellulose (Cotton, Rayon fibers) (Rayonier)

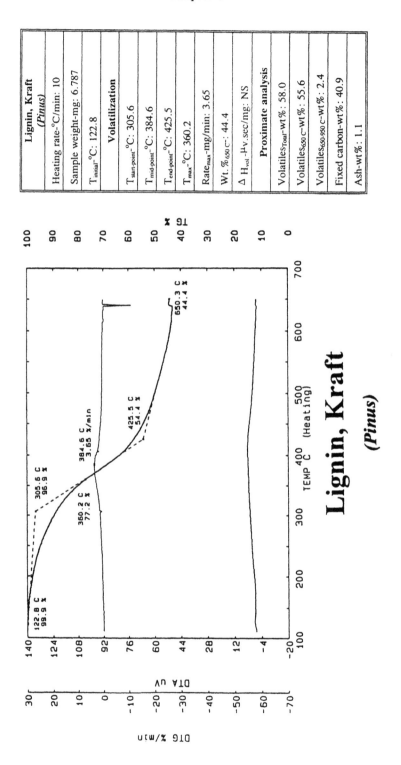

| Lignin, Kraft *(Pinus)* | |
|---|---|
| Heating rate-°C/min: 10 | |
| Sample weight-mg: 6.787 | |
| $T_{initial}$-°C: 122.8 | |
| **Volatilization** | |
| $T_{start-point}$-°C: 305.6 | |
| $T_{mid-point}$-°C: 384.6 | |
| $T_{end-point}$-°C: 425.5 | |
| $T_{max}$-°C: 360.2 | |
| $Rate_{max}$-mg/min: 3.65 | |
| $Wt.\%_{650°C}$: 44.4 | |
| $\Delta H_{vol}$-$\mu$v.sec/mg: NS | |
| **Proximate analysis** | |
| $Volatiles_{Total}$-wt%: 58.0 | |
| $Volatiles_{650°C}$-wt%: 55.6 | |
| $Volatiles_{650-950°C}$-wt%: 2.4 | |
| Fixed carbon-wt%: 40.9 | |
| Ash-wt%: 1.1 | |

# Lignin, Kraft
## *(Pinus)*

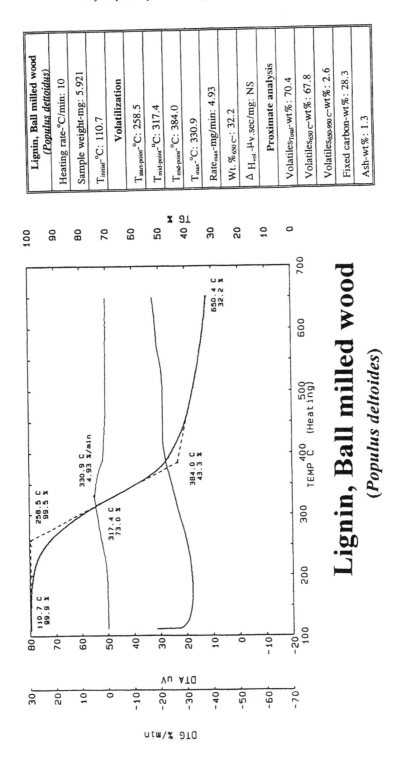

**Lignin, Ball milled wood**
*(Populus deltoides)*

| Lignin, Ball milled wood *(Populus deltoidus)* | |
|---|---|
| Heating rate-°C/min: | 10 |
| Sample weight-mg: | 5.921 |
| $T_{initial}$-°C: | 110.7 |
| **Volatilization** | |
| $T_{start-point}$-°C: | 258.5 |
| $T_{mid-point}$-°C: | 317.4 |
| $T_{end-point}$-°C: | 384.0 |
| $T_{max}$-°C: | 330.9 |
| $Rate_{max}$-mg/min: | 4.93 |
| Wt.%$_{650}$ C: | 32.2 |
| $\Delta H_{vol}$-$\mu$v.sec/mg: | NS |
| **Proximate analysis** | |
| Volatiles$_{Total}$-wt%: | 70.4 |
| Volatiles$_{650}$ C-wt%: | 67.8 |
| Volatiles$_{650-950}$ C-wt%: | 2.6 |
| Fixed carbon-wt%: | 28.3 |
| Ash-wt%: | 1.3 |

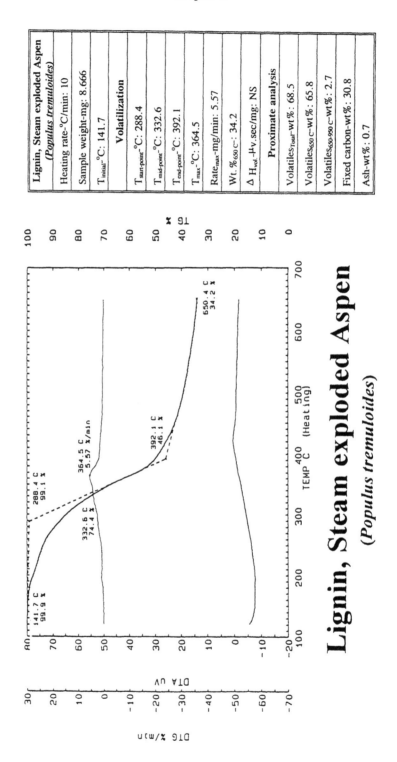

| Lignin, Steam exploded Aspen (*Populus tremuloides*) | |
| --- | --- |
| Heating rate–°C/min: | 10 |
| Sample weight–mg: | 8.666 |
| $T_{initial}$–°C: | 141.7 |
| **Volatilization** | |
| $T_{start-point}$–°C: | 288.4 |
| $T_{mid-point}$–°C: | 332.6 |
| $T_{end-point}$–°C: | 392.1 |
| $T_{max}$–°C: | 364.5 |
| $Rate_{max}$–mg/min: | 5.57 |
| Wt.%$_{650C}$: | 34.2 |
| $\Delta H_{vol}$–$\mu$v.sec/mg: | NS |
| **Proximate analysis** | |
| Volatiles$_{TToal}$–wt%: | 68.5 |
| Volatiles$_{650C}$–wt%: | 65.8 |
| Volatiles$_{650-950C}$–wt%: | 2.7 |
| Fixed carbon–wt%: | 30.8 |
| Ash–wt%: | 0.7 |

# Lignin, Steam exploded Aspen
## (*Populus tremuloides*)

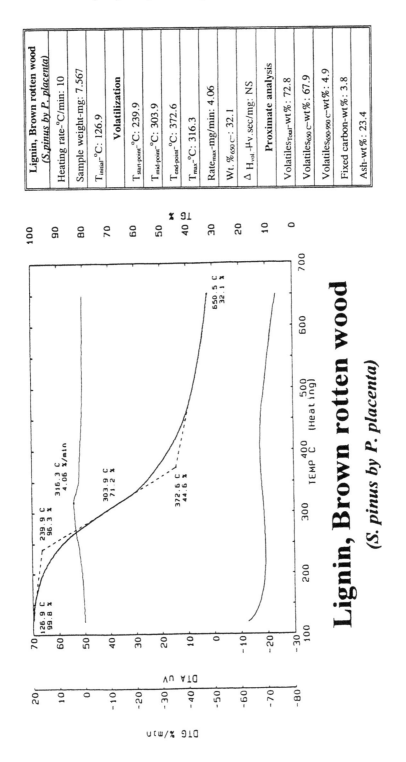

| Lignin, Brown rotten wood (*S.pinus by P. placenta*) | |
|---|---|
| Heating rate-°C/min: 10 | |
| Sample weight-mg: 7.567 | |
| $T_{initial}$-°C: 126.9 | |
| **Volatilization** | |
| $T_{start\text{-}point}$-°C: 239.9 | |
| $T_{mid\text{-}point}$-°C: 303.9 | |
| $T_{end\text{-}point}$-°C: 372.6 | |
| $T_{max}$-°C: 316.3 | |
| $Rate_{max}$-mg/min: 4.06 | |
| Wt. $\%_{650}$ c-: 32.1 | |
| $\Delta H_{vol}$-$\mu$v.sec/mg: NS | |
| **Proximate analysis** | |
| $Volatiles_{Total}$-wt%: 72.8 | |
| $Volatiles_{650 C}$-wt%: 67.9 | |
| $Volatiles_{650\text{-}950 C}$-wt%: 4.9 | |
| Fixed carbon-wt%: 3.8 | |
| Ash-wt%: 23.4 | |

# Lignin, Brown rotten wood
## (*S. pinus by P. placenta*)

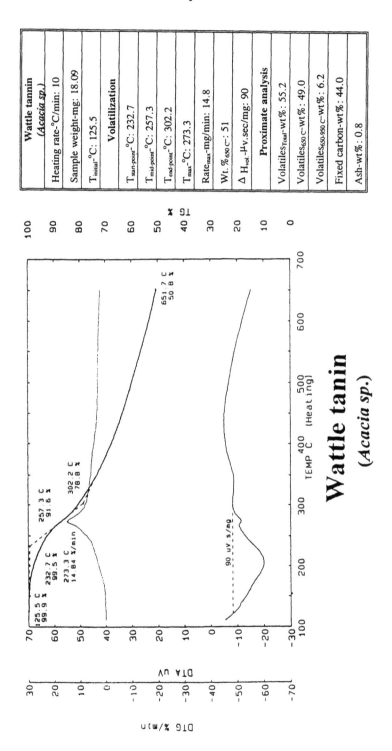

**Wattle tanin** (*Acacia sp.*)

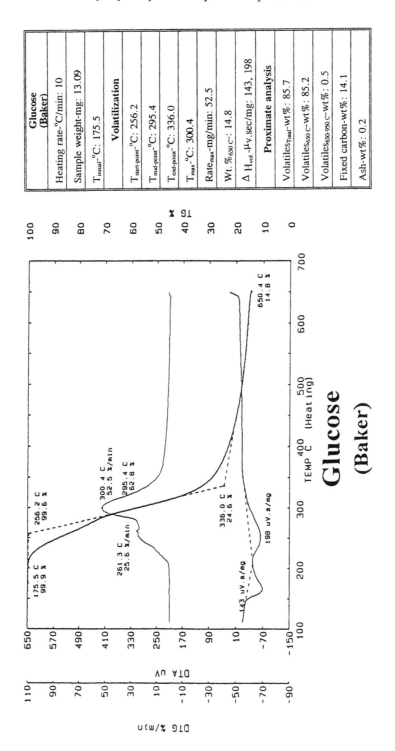

| Glucose (Baker) | |
|---|---|
| Heating rate-°C/min: | 10 |
| Sample weight-mg: | 13.09 |
| $T_{initial}$-°C: | 175.5 |
| **Volatilization** | |
| $T_{start-point}$-°C: | 256.2 |
| $T_{mid-point}$-°C: | 295.4 |
| $T_{end-point}$-°C: | 336.0 |
| $T_{max}$-°C: | 300.4 |
| $Rate_{max}$-mg/min: | 52.5 |
| $Wt.\%_{650°C}$: | 14.8 |
| $\Delta H_{vol}$-µv.sec/mg: | 143, 198 |
| **Proximate analysis** | |
| $Volatiles_{Total}$-wt%: | 85.7 |
| $Volatiles_{650°C}$-wt%: | 85.2 |
| $Volatiles_{650-950°C}$-wt%: | 0.5 |
| Fixed carbon-wt%: | 14.1 |
| Ash-wt%: | 0.2 |

# Glucose
## (Baker)

# 6

# Natural Biomass

The use of wood as a fuel source was known for many centuries around the world before the advent of fossil fuels, when wood was first replaced by coal and then subsequently by natural gas and petroleum products. In the United States and Europe, the shift from wood to fossil fuels as the major source of energy was so dramatic that the use of natural biomass as fuel dropped to a negligible value. This transition got further impetus due to the depletion of forest cover and pressure on wood resources to meet the demands of the pulp and paper industry and the logging industry. The use of wood and other natural biomass as a fuel source, however, continued in developing and undeveloped nations in Asia and Africa. In developing countries such as China and India, the dependency on wood as a fuel source was greater than 70%. In African countries, the use of natural biomass as fuel source was in some cases greater than 90%.

The embargo on the sale of petroleum-based products imposed by oil-producing nations in the Middle East in 1973 renewed interest all over the world in the use of natural biomass as an alternate source of energy. This interest was at its peak during the late 1970s and early 1980s. Research was being conducted to develop biomass-based gasifiers, combustors, and liquid fuel production. Significant progress was made in the development of biomass-based technologies during this period, but lack of better understanding of thermal behavior of biomass materials was one of several reasons why the majority of these technologies have not shown their commercial presence.

The interest in natural biomass materials as a fuel source has now been taken up from the environmental standpoint, which believes that unfavorable environmental changes such as global warming are taking place due to excessive use of fossil fuels. It is believed that if part of this fossil fuel consumption is reduced by introducing well thought out technologies based on biomass materials, then a significant reduction to the deterioration of the environment can take place. For example, biomass based-fuels are believed to keep the ecological balance of carbon dioxide in the atmosphere by recycling.

This chapter presents thermal properties for natural biomass materials such as softwoods, hardwoods, agricultural residues, and aquatic biomass samples. These properties will help in the use of these materials as alternate fuel sources.

The word *biomass* was coined in the early 1970s to refer to all living matter, but particularly with respect to use as a source of energy and fuel. It includes forest and agricultural species and even animal by-products such as manure and fats. It can refer to biomass grown specifically for energy, to biomass residues, and even to municipal waste. As discussed in the previous chapter, plants are primarily composed of cellulose, hemicellulose, and lignin plus various particular "extractives" (such as tannin). Table 6.1 shows the typical distribution of these biomass components in some natural biomass samples.

**TABLE 6.1**

Typical Composition of Some Biomass Materials

| Common name | Cellulose | Hemicellulose | Lignin | Other |
|---|---|---|---|---|
| Softwoods | 42.0 | 28 | 25–30 | 0–5 |
| Hardwoods | 42.0 | 38 | 15–20 | 0–5 |
| Flax | 71.2 | 18.6 | 2.2 | 8.0 |
| Jute | 71.6 | 13.3 | 13.1 | 2.0 |
| Hemp | 74.4 | 17.9 | 3.7 | 4.0 |
| Ramie | 76.2 | 14.6 | 0.7 | 8.5 |
| Sisal | 73.0 | 13.0 | 14.0 | 0.0 |
| Abaca | 70.2 | 21.8 | 5.7 | 2.3 |
| Cotton, crude | 95.3 | 0.0 | 0.0 | 4.7 |
| Cotton, purified | 99.9 | 0.0 | 0.0 | 0.1 |

## 6.1  WOOD

Wood is a heterogeneous material that performs three major functions in a living plant:

1. Water conduction
2. Metabolism
3. Mechanical support

It is obtained from the stems, roots, and branches of trees and shrubs and from herbaceous plants. The woody plants are seed-bearing plants known as spermatophytes. The spermatophytes are classified as gymnosperms and angiosperms.

The gymnosperms have exposed seeds that are not protected in a shell or ovary. The gymnosperms are further classified into members known as conifers, evergreen, or softwoods in general. The fruit of these plants is shaped like a cone; this is where the word *conifer* comes from. The term *softwood* has, however, found more use among the wood chemists and fuel scientists. However, one must not think that softwoods in general are softer in physical structure; they are sometimes more hard than hardwoods, which are discussed later in this chapter.

The angiosperms have an ovary that encloses the seed. The angiosperms are divided into two classes: monocotyledons and dicotyledons. Bamboo and various other palm are included in the class of monocotyledons. The dicotyledon trees are known as broad-leaved trees, deciduous trees, or hardwoods. It is important to mention again that some hardwoods are softer than certain species of softwoods.

To differentiate between the softwoods and hardwoods, it is imperative that one know some general structural details of trees and wood.

A tree is commonly divided into three parts: the crown, the stem, and the roots. The leaves and certain twigs are considered as part of the crown, which is responsible for the growth of the tree. The stem section provides the mechanical support and the metabolic functions. The roots play an important role in the life of the tree, providing nutrients to the stem through the soil.

When the tree is young, the wood that is added every year continues in sap conduction and storage of reserve food. This wood, known as *sapwood*, is physiologically active and at least a part of the tissue is alive and provides communication with the *cambium* —a dividing line between the wood and the bark. When the years pass, the inner core of the tree or the trunk grows thicker and then this wood dies and is only functional in providing the mechanical support to the tree. This is called *heartwood*. The initiation of heart-wood formation varies from one species to the other and also differs within the same species as it is dependent on many parameters such as soil, nutrition, and growth rate. Once the formation of heartwood starts in the tree, it grows towards the outside part of the tree trunk. The inner core has sapwood, which can extend to few growth rings or, in some cases, to about a hundred or more. Upon physical examination, one can differentiate between the sapwood and the heartwood by the color. The sapwood is much lighter than the heartwood.

When the tree has reached a certain stage in growth, the layer outside the cambium is referred to as bark and the one inside is referred to as wood. The outer dead bark is a collection of the cells that originally existed in the inner living bark. The inner bark, which is narrow, contains conducting and storage cells.

## 6.1.1  Softwoods and Hardwoods

Conifers or softwood cells have their longest axis oriented either longitudinally or radially. The cell type that has the longest axis to the grain is the tracheid, which is an extremely long linear cell about 60–70 times longer than its diameter. The second type of these kind of cells are the epithelial cells surrounding the longitu-dinal resin canals.

The hardwoods differ greatly from softwoods because they con-tain vessels known as pores, which have little or no radial align-ment of the longitudinal cells. The term *porous* is used to describe the presence of pores in the wood of the dicotyledonous species. In contrast, the conifers or the softwoods do not possess such a feature and are termed *nonporous*. Additional longitudinal cells that may occur in the hardwoods and contribute to their complexity are tra-cheids, fibers, and longitudinal parenchyma.

The tracheids in the hardwoods are of two types: the vascular tracheid and the vasicentric tracheid. The former resembles the in-dividual member of small vessels and does not have perforations, while the latter resembles short conifers. This does not occur in all types of species.

The fibers in hardwoods are of two types: the fiber tracheids and the libriform fibers. Both of these types of fibrous cells are like any other fibrous cells but they are less pitted or have less markings when compared with those obtained from softwoods. This is pri-marily due to the fact that in hardwoods the fibers are required only to support the cells and they do not help in transporting liq-uids, while in softwoods these cells have to perform both the functions. The average length of fibers in hardwoods is anywhere from 0.8 to 1.9 mm, while for softwoods this length is between 3 and 6 mm.

The longitudinal parenchyma in porous woods is strand parenchyma, in which the short cells are arranged in rows along the grain of the wood. The strands of longitudinal parenchyma are shorter in hardwoods than in softwoods.

## 6.1.2 Physical Properties of Wood

*Specific Gravity*

Specific gravity of wood is the ratio between the oven-dry weight of wood and the weight of an equal volume of water at a specified temperature. Most of the woods have specific gravity of 0.35 to 0.65 on a green volume basis. Some scientists refer to this specific gravity as the apparent specific gravity of wood because the specific gravity of the cell wall substance itself is in the neighborhood of 1.53. If all the air trapped in the cavity of the wood can be replaced with water, then the specific gravity of wood would be greater than 1.0 and they would tend to sink in water.

*Moisture Content*

Wood has a high affinity towards water under oven-dry conditions and continues to decrease with the increase in humidity. The saturation point of most woods is reached at about 30% moisture content on a weight basis. After the saturation point, the water attached to woods surface is called *free water* and has little effect on the physical or chemical properties of wood. However, the water that is absorbed in the cell walls changes the physical properties of wood. The most prominent among them is that dry wood tends to swell upon contact with water.

The determination of the moisture content of wood is also essential from the energy point of view to accurately determine the net heating value. In most thermal conversion processes, the inaccurate determination of the wood moisture content leads to a drop in the estimated thermal efficiency of the process unit.

## 6.2  AGRICULTURAL RESIDUES

Agricultural residue is any material from crop plants that is left over after the desired portion of the plant has been utilized. Some common examples of agricultural residues include the straws, hulls, seeds, linters, and other similar by-products of agriculture. These

**TABLE 6.2**

Annual Availability of Crop Residues

| Residue category | Million dry tons/year |
|---|---|
| Corn and sorghum (field) | 96.6 |
| Small grains and grasses (field) | 131.8 |
| Other crops | 42.3 |
| Collected residues | 7.3 |
| Total | 278.0 |

*Source*: Inman (1981).

leftover parts in most cases are left in the field after harvest and are replowed into the soil. The agricultural residues have shown promise as a fuel source in gasifiers and pyrolysers. The energy content of these materials lies in the range of 2000–4000 kcal/g and they are widely available around the world at low or negative (on site) cost. However, one of the major disadvantages of these residues from the energy point of view lies in the fact that they have a relatively high ash content (2–7%) and in some cases, like rice hull and rice straw, the ash content is as high as 20% and 10%, respectively. This is in contrast to wood, in which the typical ash content is 0.1–2% on a weight basis. The annual availability of certain crop residues is shown in Table 6.2.

The energy contribution from biomass materials in the United States has been estimated at 20% and this number has not changed very much over the past several years. The distribution of various energy sources among the different biomass forms are shown in Table 6.3. The data presented in this book are presented on a mois-

**TABLE 6.3**

Biomass Resources as Energy in the United States

| Resource | $10^6$ dry tons/year | Quads/year[a] |
|---|---|---|
| Crop residues | 278.0 | 4.15 |
| Animal manures | 26.5 | 0.33 |
| Unused mill residues | 24.1 | 0.41 |
| Logging residues | 83.2 | 1.41 |
| Municipal solid wastes | 130.0 | 1.63 |
| Standing forests | 384.0 | 6.51 |
| Total | 925.8 | 14.44 |

[a] 1 quad = $10^{15}$ Btu/year.
*Source*: Inman (1981).

ture-free basis as every sample was dried at 120°C before each test run.

## 6.3   AQUATIC BIOMASS

Aquatic biomass is considered a potential source of energy. It grows in the oceans (which cover more than three-fourths of the globe) and in lakes, streams, and rivers. There have been many proposals to collect and use various forms of aquatic biomass for energy, but none have yet been commercially used. Indications are that if these resources are properly tapped, then algae can turn out to be a major oil (triglyceride) producing source. The organic carbon content of most algae is around 50–55% in contrast to most terrestrial plants, which contain about 45% organic carbon content. However, this apparent advantage of aquatic biomass materials is overshadowed by the presence of high ash content (25–35%). In addition, they also have a high water content, which nullifies most of the energy content present in these materials.

## 6.4   THERMAL DEGRADATION OF NATURAL BIOMASS MATERIALS

Since Table 6.1 shows that there is little difference between the cellulose, hemicellulose, and lignin (C/H/L) composition of hardwood and softwood, it is not surprising that their thermograms are quite similar, and appear to be the arithmetic sums of the thermal degradation pattern of these individual components (C/H/L) discussed in Chapter 5. This additivity of components is experimentally demonstrated in a thermogram for one deashed wood sample (*Populus deltoides*) presented in Chapter 12. This phenomenon of component additivity is also found to be true for many of the agricultural specimens. Hence, one may even expect that the thermograms for a majority of natural biomass samples could be deconvoluted to give the proportions of C/H/L in a specific sample. However, since ash plays a major catalytic role in determining the volatility pattern of most naturally occurring biomass samples, this process of component deconvolution is not very successful for practical use, because it would have to be attempted after the tedious process of ash removal.

The difference in the thermograms obtained for wood and agricultural samples is highlighted by the percentage distribution of fixed carbon and ash content. It is clearly seen from the thermograms presented in this chapter that as a class, agricultural materials contain more ash and produce more fixed carbon than woody materials. This suggests that agricultural residues could be a useful source of charcoal production, with briquetting for fuel use in developing countries. This would help decrease the deforestation pattern in these countries.

The aquatic specimens are unique in having a sudden loss of mass at about 120°C. Since all the specimens are dried equally at 110°C before running the thermograms, this is puzzling. Perhaps there are small, impermeable cells containing water that explode at temperatures higher than 120°C and result in sudden weight loss due to the evaporation of the entrapped water content.

# SOFTWOODS

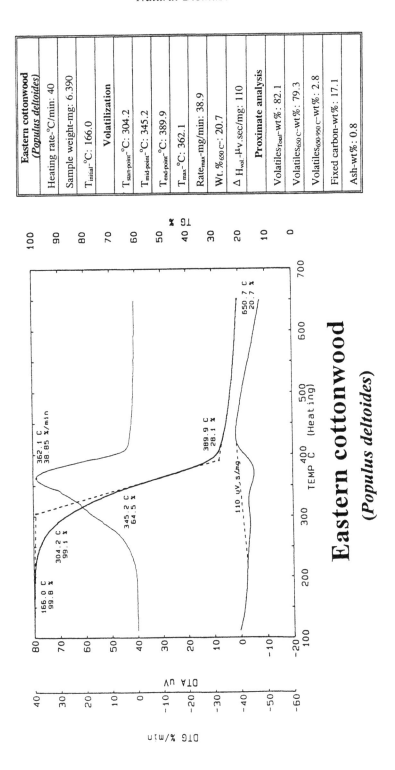

| Eastern cottonwood (*Populus deltoides*) | |
|---|---|
| Heating rate-°C/min: | 40 |
| Sample weight-mg: | 6.390 |
| $T_{initial}$-°C: | 166.0 |
| **Volatilization** | |
| $T_{start-point}$-°C: | 304.2 |
| $T_{mid-point}$-°C: | 345.2 |
| $T_{end-point}$-°C: | 389.9 |
| $T_{max}$-°C: | 362.1 |
| $Rate_{max}$-mg/min: | 38.9 |
| Wt.%$_{650°C}$: | 20.7 |
| $\Delta H_{vol}$-μv.sec/mg: | 110 |
| **Proximate analysis** | |
| Volatiles$_{Total}$-wt%: | 82.1 |
| Volatiles$_{650C}$-wt%: | 79.3 |
| Volatiles$_{650-950C}$-wt%: | 2.8 |
| Fixed carbon-wt%: | 17.1 |
| Ash-wt%: | 0.8 |

# Eastern cottonwood
## (*Populus deltoides*)

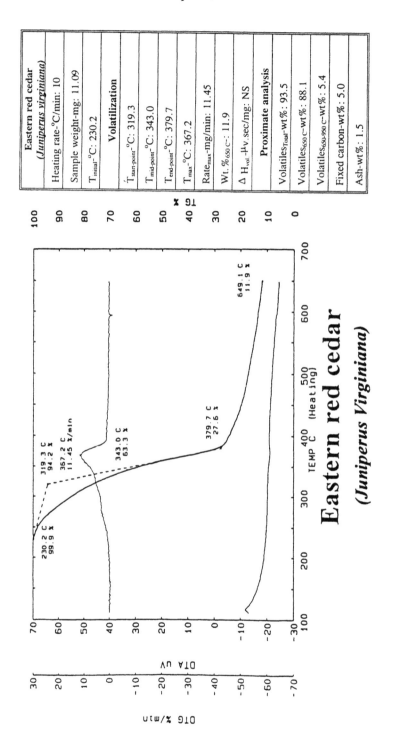

| Eastern red cedar (*Juniperus virginiana*) | |
| --- | --- |
| Heating rate-°C/min: | 10 |
| Sample weight-mg: | 11.09 |
| $T_{initial}$-°C: | 230.2 |
| **Volatilization** | |
| $T_{start-point}$-°C: | 319.3 |
| $T_{mid-point}$-°C: | 343.0 |
| $T_{end-point}$-°C: | 379.7 |
| $T_{max}$-°C: | 367.2 |
| $Rate_{max}$-mg/min: | 11.45 |
| $Wt.\%_{650\,C}$: | 11.9 |
| $\Delta H_{vol}$-µv.sec/mg: | NS |
| **Proximate analysis** | |
| $Volatiles_{T total}$-wt%: | 93.5 |
| $Volatiles_{650\,C}$-wt%: | 88.1 |
| $Volatiles_{650-950\,C}$-wt%: | 5.4 |
| Fixed carbon-wt%: | 5.0 |
| Ash-wt%: | 1.5 |

# Eastern red cedar
## (*Juniperus Virginiana*)

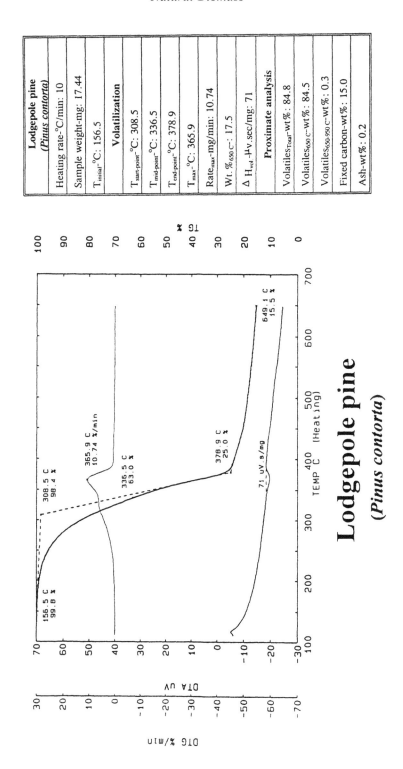

| Lodgepole pine *(Pinus contorta)* | |
|---|---|
| Heating rate-°C/min: 10 | |
| Sample weight-mg: 17.44 | |
| $T_{initial}$-°C: 156.5 | |
| **Volatilization** | |
| $T_{start-point}$-°C: 308.5 | |
| $T_{mid-point}$-°C: 336.5 | |
| $T_{end-point}$-°C: 378.9 | |
| $T_{max}$-°C: 365.9 | |
| $Rate_{max}$-mg/min: 10.74 | |
| Wt. %$_{650 C}$: 17.5 | |
| $\Delta H_{vol}$ -μv.sec/mg: 71 | |
| **Proximate analysis** | |
| Volatiles$_{Total}$-wt%: 84.8 | |
| Volatiles$_{650 C}$-wt%: 84.5 | |
| Volatiles$_{650-950 C}$-wt%: 0.3 | |
| Fixed carbon-wt%: 15.0 | |
| Ash-wt%: 0.2 | |

# Lodgepole pine
*(Pinus contorta)*

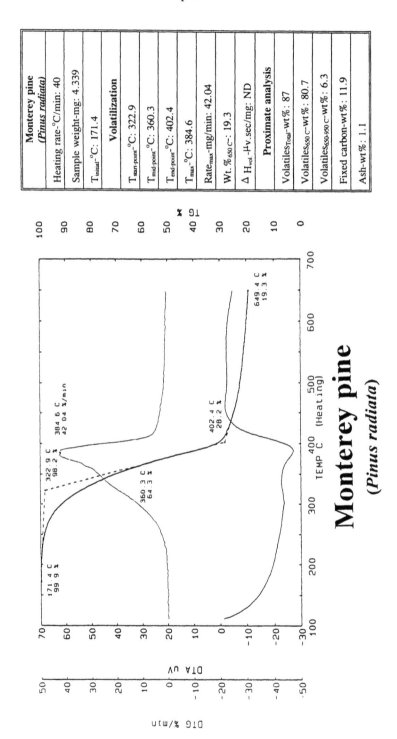

| Monterey pine (Pinus radiata) | |
|---|---|
| Heating rate-°C/min: | 40 |
| Sample weight-mg: | 4.339 |
| $T_{initial}$-°C: | 171.4 |
| **Volatilization** | |
| $T_{start-point}$-°C: | 322.9 |
| $T_{mid-point}$-°C: | 360.3 |
| $T_{end-point}$-°C: | 402.4 |
| $T_{max}$-°C: | 384.6 |
| $Rate_{max}$-mg/min: | 42.04 |
| $Wt. \%_{650°C}$: | 19.3 |
| $\Delta H_{vol}$-µv.sec/mg: | ND |
| **Proximate analysis** | |
| $Volatiles_{Total}$-wt%: | 87 |
| $Volatiles_{650°C}$-wt%: | 80.7 |
| $Volatiles_{650-950°C}$-wt%: | 6.3 |
| Fixed carbon-wt%: | 11.9 |
| Ash-wt%: | 1.1 |

# Monterey pine
## (*Pinus radiata*)

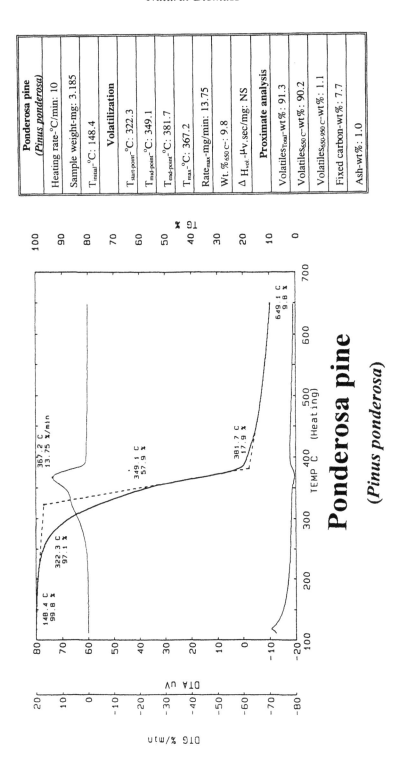

| Ponderosa pine |
| (*Pinus ponderosa*) |
| Heating rate-°C/min: 10 |
| Sample weight-mg: 3.185 |
| $T_{initial}$-°C: 148.4 |
| **Volatilization** |
| $T_{start-point}$-°C: 322.3 |
| $T_{mid-point}$-°C: 349.1 |
| $T_{end-point}$-°C: 381.7 |
| $T_{max}$-°C: 367.2 |
| Rate$_{max}$-mg/min: 13.75 |
| Wt.%$_{650}$ c-: 9.8 |
| $\Delta$ H$_{vol.}$-$\mu$v.sec/mg: NS |
| **Proximate analysis** |
| Volatiles$_{Total}$-wt%: 91.3 |
| Volatiles$_{650\,c}$-wt%: 90.2 |
| Volatiles$_{650-950\,c}$-wt%: 1.1 |
| Fixed carbon-wt%: 7.7 |
| Ash-wt%: 1.0 |

# Ponderosa pine

(*Pinus ponderosa*)

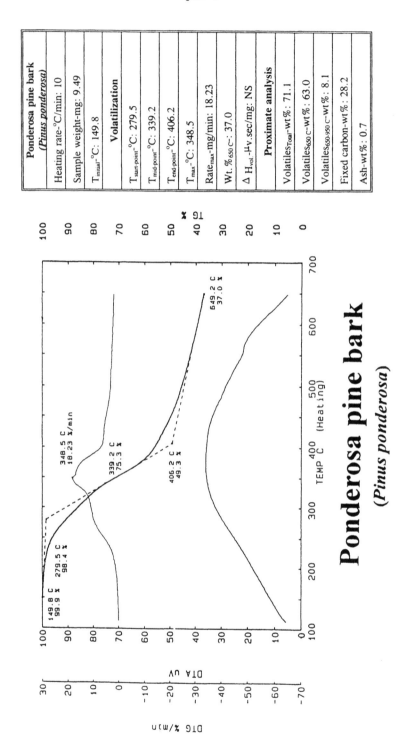

| Ponderosa pine bark (*Pinus ponderosa*) | |
|---|---|
| Heating rate-°C/min: | 10 |
| Sample weight-mg: | 9.49 |
| $T_{initial}$-°C: | 149.8 |
| **Volatilization** | |
| $T_{start-point}$-°C: | 279.5 |
| $T_{mid-point}$-°C: | 339.2 |
| $T_{end-point}$-°C: | 406.2 |
| $T_{max}$-°C: | 348.5 |
| $Rate_{max}$-mg/min: | 18.23 |
| Wt.$\%_{650C}$: | 37.0 |
| $\Delta H_{vol.}$-$^{14}$v.sec/mg: | NS |
| **Proximate analysis** | |
| Volatiles$_{Total}$-wt%: | 71.1 |
| Volatiles$_{650C}$-wt%: | 63.0 |
| Volatiles$_{650-950C}$-wt%: | 8.1 |
| Fixed carbon-wt%: | 28.2 |
| Ash-wt%: | 0.7 |

# Ponderosa pine bark
## (*Pinus ponderosa*)

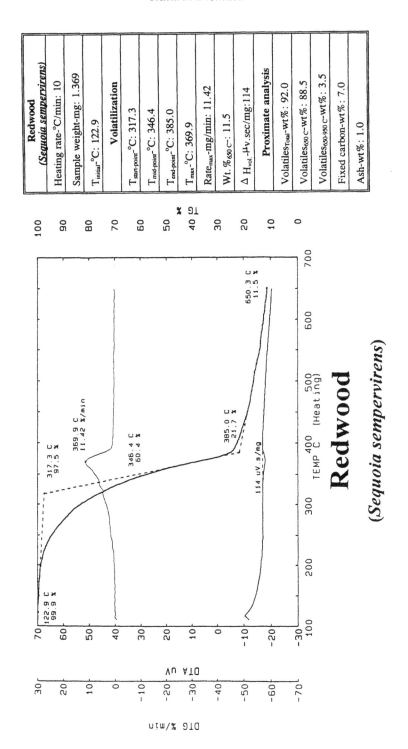

| Redwood (*Sequoia sempervirens*) | |
|---|---|
| Heating rate–°C/min: | 10 |
| Sample weight–mg: | 1.369 |
| $T_{initial}$–°C: | 122.9 |
| **Volatilization** | |
| $T_{start-point}$–°C: | 317.3 |
| $T_{mid-point}$–°C: | 346.4 |
| $T_{end-point}$–°C: | 385.0 |
| $T_{max}$–°C: | 369.9 |
| $Rate_{max}$–mg/min: | 11.42 |
| Wt.%$_{650°C}$: | 11.5 |
| $\Delta H_{vol}$–μv.sec/mg: | 114 |
| **Proximate analysis** | |
| Volatiles$_{Total}$–wt%: | 92.0 |
| Volatiles$_{650°C}$–wt%: | 88.5 |
| Volatiles$_{650-950°C}$–wt%: | 3.5 |
| Fixed carbon–wt%: | 7.0 |
| Ash–wt%: | 1.0 |

**Redwood**

(*Sequoia sempervirens*)

117

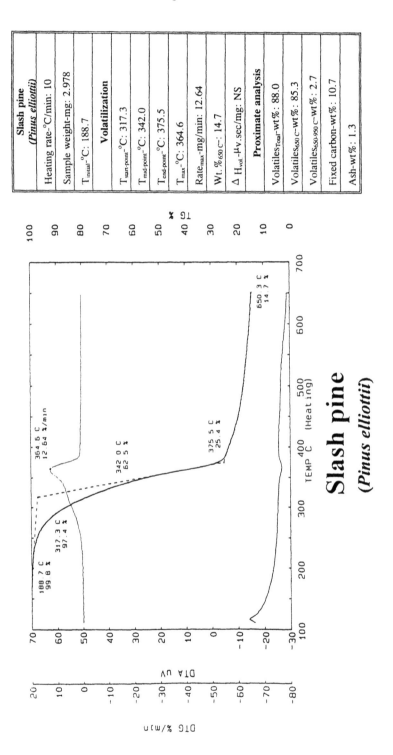

| Slash pine (Pinus elliottii) | |
|---|---|
| Heating rate-°C/min: 10 | |
| Sample weight-mg: 2.978 | |
| $T_{initial}$-°C: 188.7 | |
| **Volatilization** | |
| $T_{start-point}$-°C: 317.3 | |
| $T_{mid-point}$-°C: 342.0 | |
| $T_{end-point}$-°C: 375.5 | |
| $T_{max}$-°C: 364.6 | |
| $Rate_{max}$-mg/min: 12.64 | |
| Wt.%$_{650C}$: 14.7 | |
| $\Delta H_{vol}$-μv.sec/mg: NS | |
| **Proximate analysis** | |
| Volatiles$_{Total}$-wt%: 88.0 | |
| Volatiles$_{650C}$-wt%: 85.3 | |
| Volatiles$_{650-990C}$-wt%: 2.7 | |
| Fixed carbon-wt%: 10.7 | |
| Ash-wt%: 1.3 | |

## Slash pine
### (Pinus elliottii)

| Sugar pine (*Pinus lambertiana*) | |
|---|---|
| Heating rate-°C/min: | 10 |
| Sample weight-mg: | 1.310 |
| $T_{initial}$-°C: | 122.9 |
| **Volatilization** | |
| $T_{start-point}$-°C: | 308.7 |
| $T_{mid-point}$-°C: | 340.9 |
| $T_{end-point}$-°C: | 383.4 |
| $T_{max}$-°C: | 367.1 |
| $Rate_{max}$-mg/min: | 11.11 |
| Wt.%$_{650C}$: | 5.5 |
| $\Delta H_{vol.}$-$\mu$v.sec/mg: | NS |
| **Proximate analysis** | |
| Volatiles$_{Total}$-wt%: | 98.1 |
| Volatiles$_{650C}$-wt%: | 94.5 |
| Volatiles$_{650-950C}$-wt%: | 3.6 |
| Fixed carbon-wt%: | 3.1 |
| Ash-wt%: | 0.5 |

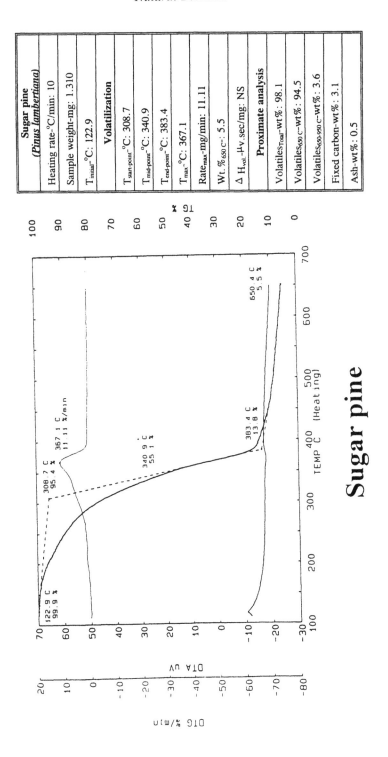

**Sugar pine**

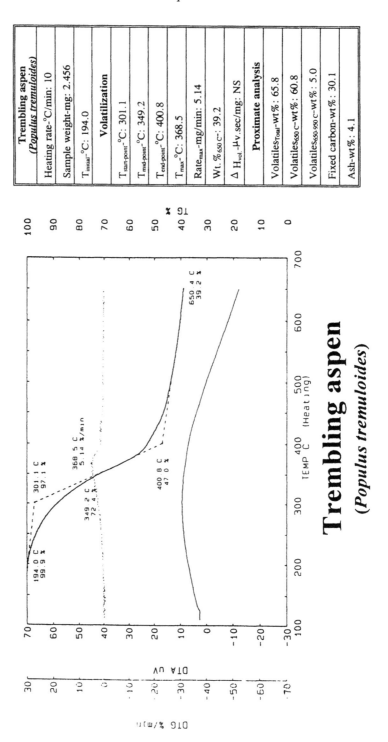

| Trembling aspen (*Populus tremuloides*) | |
|---|---|
| Heating rate-°C/min: 10 | |
| Sample weight-mg: 2.456 | |
| $T_{initial}$-°C: 194.0 | |
| **Volatilization** | |
| $T_{start-point}$-°C: 301.1 | |
| $T_{mid-point}$-°C: 349.2 | |
| $T_{end-point}$-°C: 400.8 | |
| $T_{max}$-°C: 368.5 | |
| $Rate_{max}$-mg/min: 5.14 | |
| Wt.$\%_{650°C}$: 39.2 | |
| $\Delta H_{vol.}$-$\mu$v.sec/mg: NS | |
| **Proximate analysis** | |
| Volatiles$_{Total}$-wt%: 65.8 | |
| Volatiles$_{650°C}$-wt%: 60.8 | |
| Volatiles$_{650-950°C}$-wt%: 5.0 | |
| Fixed carbon-wt%: 30.1 | |
| Ash-wt%: 4.1 | |

# Trembling aspen
## (*Populus tremuloides*)

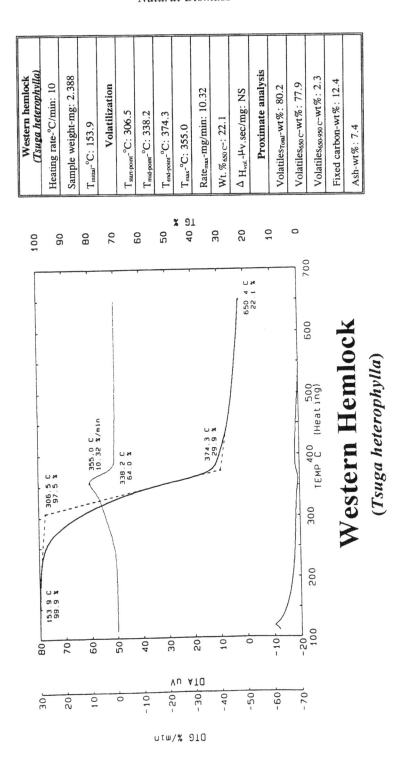

| Western hemlock (*Tsuga heterophylla*) | |
|---|---|
| Heating rate-°C/min: 10 | |
| Sample weight-mg: 2.388 | |
| $T_{initial}$-°C: 153.9 | |
| **Volatilization** | |
| $T_{start-point}$-°C: 306.5 | |
| $T_{mid-point}$-°C: 338.2 | |
| $T_{end-point}$-°C: 374.3 | |
| $T_{max}$-°C: 355.0 | |
| $Rate_{max}$-mg/min: 10.32 | |
| Wt.%$_{650}$ C: 22.1 | |
| $\Delta H_{vol}$-$\mu$v.sec/mg: NS | |
| **Proximate analysis** | |
| Volatiles$_{Total}$-wt%: 80.2 | |
| Volatiles$_{650}$ C-wt%: 77.9 | |
| Volatiles$_{650-950}$ C-wt%: 2.3 | |
| Fixed carbon-wt%: 12.4 | |
| Ash-wt%: 7.4 | |

# Western Hemlock
## (*Tsuga heterophylla*)

121

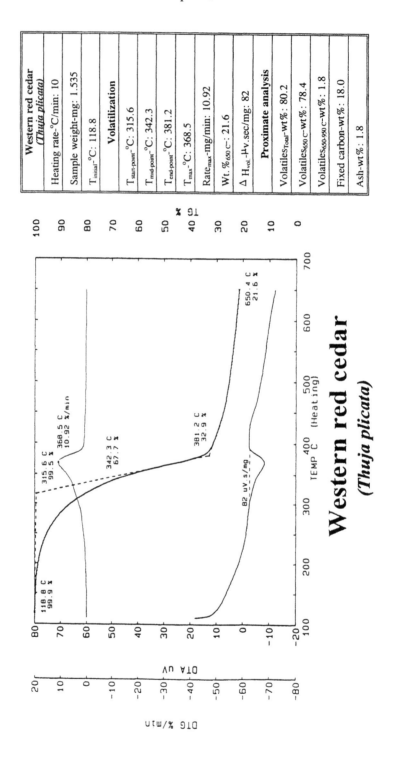

| Western red cedar (*Thuja plicata*) | |
|---|---|
| Heating rate-°C/min: | 10 |
| Sample weight-mg: | 1.535 |
| $T_{initial}$-°C: | 118.8 |
| **Volatilization** | |
| $T_{start-point}$-°C: | 315.6 |
| $T_{mid-point}$-°C: | 342.3 |
| $T_{end-point}$-°C: | 381.2 |
| $T_{max}$-°C: | 368.5 |
| $Rate_{max}$-mg/min: | 10.92 |
| Wt.%$_{650 C}$: | 21.6 |
| $\Delta H_{vol}$.-$\mu$v.sec/mg: | 82 |
| **Proximate analysis** | |
| Volatiles$_{Total}$-wt%: | 80.2 |
| Volatiles$_{650 C}$-wt%: | 78.4 |
| Volatiles$_{650-950 C}$-wt%: | 1.8 |
| Fixed carbon-wt%: | 18.0 |
| Ash-wt%: | 1.8 |

**Western red cedar**
*(Thuja plicata)*

# HARDWOODS

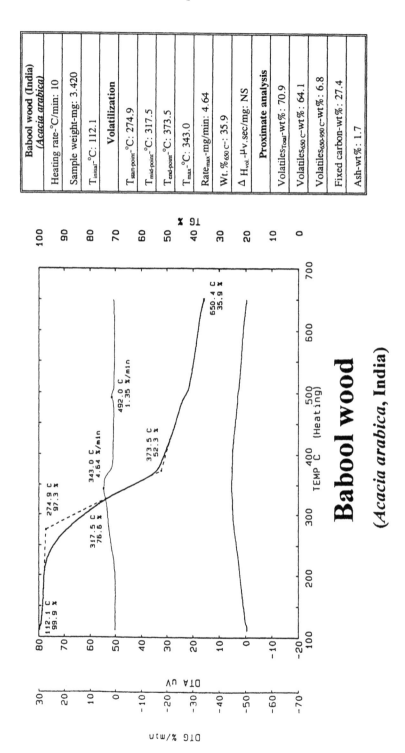

| Babool wood (India) |
|---|
| *(Acacia arabica)* |
| Heating rate-°C/min: 10 |
| Sample weight-mg: 3.420 |
| $T_{initial}$-°C: 112.1 |
| **Volatilization** |
| $T_{start-point}$-°C: 274.9 |
| $T_{mid-point}$-°C: 317.5 |
| $T_{end-point}$-°C: 373.5 |
| $T_{max}$-°C: 343.0 |
| $Rate_{max}$-mg/min: 4.64 |
| Wt.%$_{650C}$: 35.9 |
| $\Delta H_{vol}$-$\mu$v.sec/mg: NS |
| **Proximate analysis** |
| Volatiles$_{Total}$-wt%: 70.9 |
| Volatiles$_{650C}$-wt%: 64.1 |
| Volatiles$_{650-990C}$-wt%: 6.8 |
| Fixed carbon-wt%: 27.4 |
| Ash-wt%: 1.7 |

# Babool wood

### (*Acacia arabica*, India)

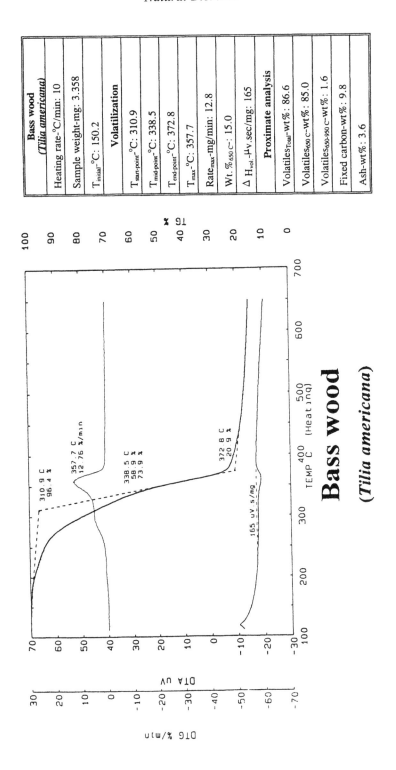

| Bass wood (*Tilia americana*) |
| --- |
| Heating rate-°C/min: 10 |
| Sample weight-mg: 3.358 |
| $T_{initial}$-°C: 150.2 |
| **Volatilization** |
| $T_{start-point}$-°C: 310.9 |
| $T_{mid-point}$-°C: 338.5 |
| $T_{end-point}$-°C: 372.8 |
| $T_{max}$-°C: 357.7 |
| $Rate_{max}$-mg/min: 12.8 |
| Wt.%$_{650C}$: 15.0 |
| $\Delta H_{vol.}$-μv.sec/mg: 165 |
| **Proximate analysis** |
| Volatiles$_{Tcoal}$-wt%: 86.6 |
| Volatiles$_{650C}$-wt%: 85.0 |
| Volatiles$_{650-950C}$-wt%: 1.6 |
| Fixed carbon-wt%: 9.8 |
| Ash-wt%: 3.6 |

**Bass wood**
**(*Tilia americana*)**

310.9 C
96.4 %

357.7 C
12.76 %/min

338.5 C
58.9 %
73.9 %

372.8 C
20.9 %

165 uv.s/mg

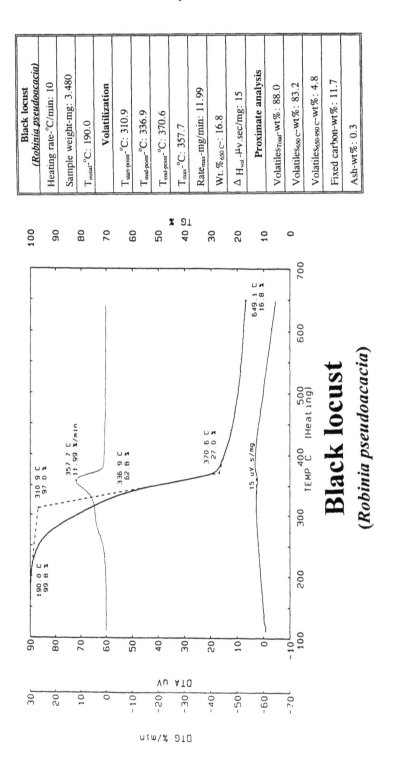

| Black locust (*Robinia pseudoacacia*) | |
|---|---|
| Heating rate-°C/min: 10 | |
| Sample weight-mg: 3.480 | |
| $T_{initial}$-°C: 190.0 | |
| **Volatilization** | |
| $T_{start-point}$-°C: 310.9 | |
| $T_{mid-point}$-°C: 336.9 | |
| $T_{end-point}$-°C: 370.6 | |
| $T_{max}$-°C: 357.7 | |
| $Rate_{max}$-mg/min: 11.99 | |
| Wt.%$_{650C}$: 16.8 | |
| $\Delta H_{vol}$-$\mu$v.sec/mg: 15 | |
| **Proximate analysis** | |
| Volatiles$_{Total}$-wt%: 88.0 | |
| Volatiles$_{650C}$-wt%: 83.2 | |
| Volatiles$_{650-950C}$-wt%: 4.8 | |
| Fixed carbon-wt%: 11.7 | |
| Ash-wt%: 0.3 | |

# Black locust
## (*Robinia pseudoacacia*)

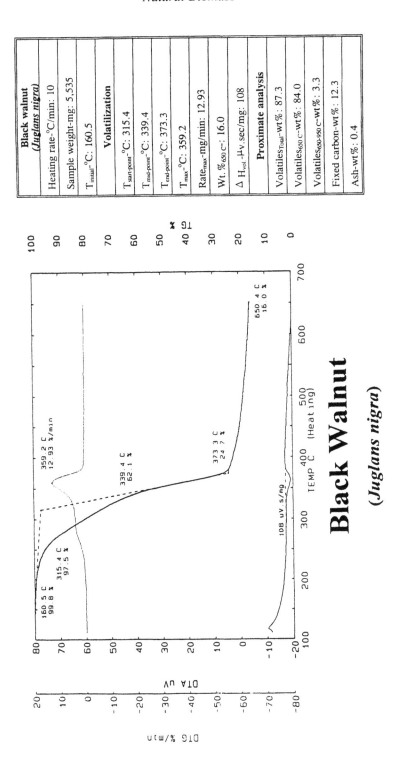

| Black walnut (*Juglans nigra*) | |
|---|---|
| Heating rate-°C/min: 10 | |
| Sample weight-mg: 5,535 | |
| $T_{initial}$-°C: 160.5 | |
| **Volatilization** | |
| $T_{start-point}$-°C: 315.4 | |
| $T_{mid-point}$-°C: 339.4 | |
| $T_{end-point}$-°C: 373.3 | |
| $T_{max}$-°C: 359.2 | |
| $Rate_{max}$-mg/min: 12.93 | |
| Wt.$\%_{650 c}$: 16.0 | |
| $\Delta H_{vol}$-μv.sec/mg: 108 | |
| **Proximate analysis** | |
| Volatiles$_{Total}$-wt%: 87.3 | |
| Volatiles$_{650 c}$-wt%: 84.0 | |
| Volatiles$_{650-950 c}$-wt%: 3.3 | |
| Fixed carbon-wt%: 12.3 | |
| Ash-wt%: 0.4 | |

# Black Walnut
## (*Juglans nigra*)

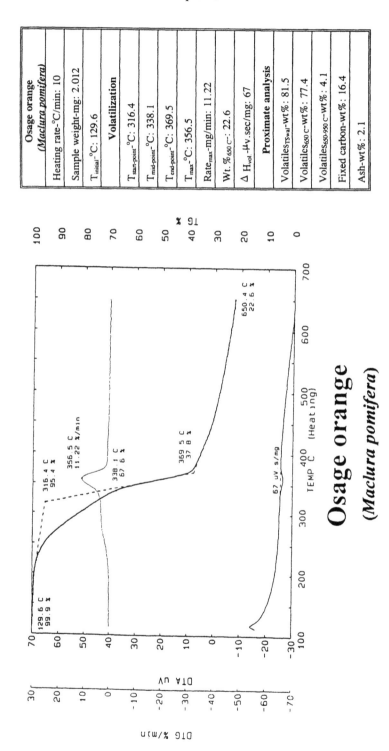

| Osage orange *(Maclura pomifera)* |
|---|
| Heating rate-°C/min: 10 |
| Sample weight-mg: 2.012 |
| $T_{initial}$-°C: 129.6 |
| **Volatilization** |
| $T_{start-point}$-°C: 316.4 |
| $T_{mid-point}$-°C: 338.1 |
| $T_{end-point}$-°C: 369.5 |
| $T_{max}$-°C: 356.5 |
| $Rate_{max}$-mg/min: 11.22 |
| Wt.%$_{650 C}$: 22.6 |
| $\Delta H_{vol.}$-μv.sec/mg: 67 |
| **Proximate analysis** |
| Volatiles$_{TSwal}$-wt%: 81.5 |
| Volatiles$_{650 C}$-wt%: 77.4 |
| Volatiles$_{650-950 C}$-wt%: 4.1 |
| Fixed carbon-wt%: 16.4 |
| Ash-wt%: 2.1 |

Osage orange
*(Maclura pomifera)*

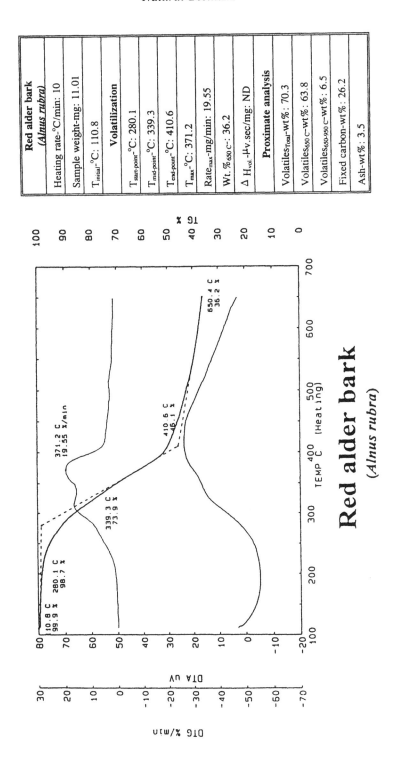

| Red alder bark *(Alnus rubra)* | |
| --- | --- |
| Heating rate-°C/min: 10 | |
| Sample weight-mg: 11.01 | |
| $T_{initial}$-°C: 110.8 | |
| **Volatilization** | |
| $T_{start-point}$-°C: 280.1 | |
| $T_{mid-point}$-°C: 339.3 | |
| $T_{end-point}$-°C: 410.6 | |
| $T_{max}$-°C: 371.2 | |
| $Rate_{max}$-mg/min: 19.55 | |
| Wt.%$_{650 C}$: 36.2 | |
| $\Delta H_{vol}$-$\mu$v.sec/mg: ND | |
| **Proximate analysis** | |
| Volatiles$_{Total}$-wt%: 70.3 | |
| Volatiles$_{650 C}$-wt%: 63.8 | |
| Volatiles$_{650-950 C}$-wt%: 6.5 | |
| Fixed carbon-wt%: 26.2 | |
| Ash-wt%: 3.5 | |

## Red alder bark
*(Alnus rubra)*

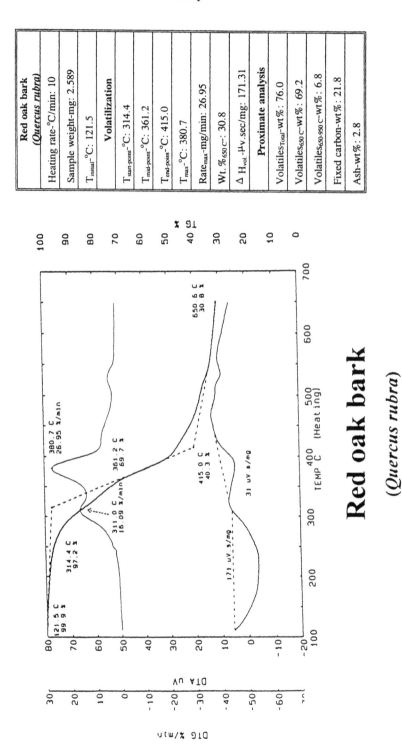

| Red oak bark (*Quercus rubra*) | |
|---|---|
| Heating rate-°C/min: 10 | |
| Sample weight-mg: 2.589 | |
| $T_{initial}$-°C: 121.5 | |
| **Volatilization** | |
| $T_{start-point}$-°C: 314.4 | |
| $T_{mid-point}$-°C: 361.2 | |
| $T_{end-point}$-°C: 415.0 | |
| $T_{max}$-°C: 380.7 | |
| $Rate_{max}$-mg/min: 26.95 | |
| Wt. $\%_{650 °C}$: 30.8 | |
| $\Delta H_{vol.}$-$\mu$v.sec/mg: 171.31 | |
| **Proximate analysis** | |
| Volatiles$_{Total}$-wt%: 76.0 | |
| Volatiles$_{650 °C}$-wt%: 69.2 | |
| Volatiles$_{650-950 °C}$-wt%: 6.8 | |
| Fixed carbon-wt%: 21.8 | |
| Ash-wt%: 2.8 | |

# Red oak bark
## (*Quercus rubra*)

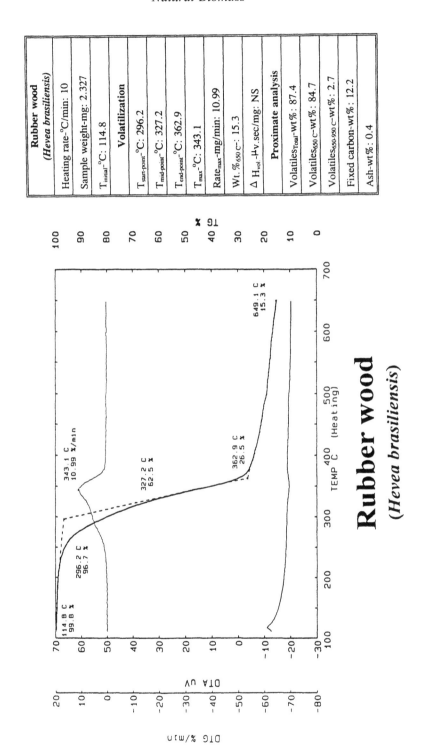

| Rubber wood (*Hevea brasiliensis*) | |
|---|---|
| Heating rate–°C/min: 10 | |
| Sample weight–mg: 2.327 | |
| $T_{initial}$–°C: 114.8 | |
| **Volatilization** | |
| $T_{start-point}$–°C: 296.2 | |
| $T_{mid-point}$–°C: 327.2 | |
| $T_{end-point}$–°C: 362.9 | |
| $T_{max}$–°C: 343.1 | |
| $Rate_{max}$–mg/min: 10.99 | |
| Wt.$\%_{650 C}$–: 15.3 | |
| $\Delta H_{vol.}$–$\mu$v.sec/mg: NS | |
| **Proximate analysis** | |
| $Volatiles_{Total}$–wt%: 87.4 | |
| $Volatiles_{650 C}$–wt%: 84.7 | |
| $Volatiles_{650-950 C}$–wt%: 2.7 | |
| Fixed carbon–wt%: 12.2 | |
| Ash–wt%: 0.4 | |

**Rubber wood**
(*Hevea brasiliensis*)

# AGRICULTURAL RESIDUES

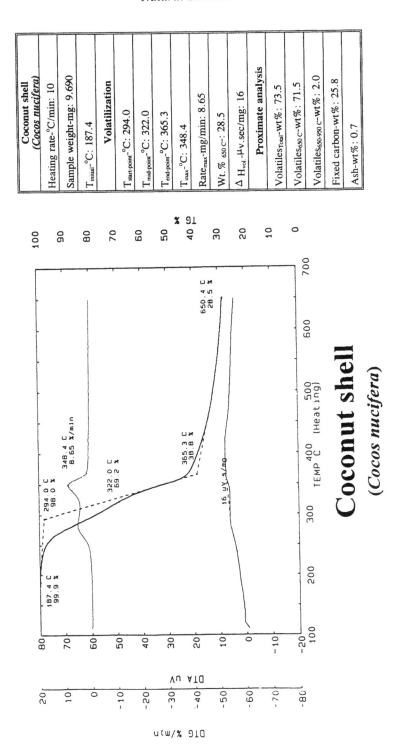

| Coconut shell (*Cocos nucifera*) | |
|---|---|
| Heating rate-°C/min: | 10 |
| Sample weight-mg: | 9.690 |
| $T_{initial}$-°C: | 187.4 |
| **Volatilization** | |
| $T_{start-point}$-°C: | 294.0 |
| $T_{mid-point}$-°C: | 322.0 |
| $T_{end-point}$-°C: | 365.3 |
| $T_{max}$-°C: | 348.4 |
| $Rate_{max}$-mg/min: | 8.65 |
| Wt. % $_{650 C}$: | 28.5 |
| $\Delta H_{vol}$-$\mu$v.sec/mg: | 16 |
| **Proximate analysis** | |
| Volatiles$_{Total}$-wt%: | 73.5 |
| Volatiles$_{650 C}$-wt%: | 71.5 |
| Volatiles$_{650-950 C}$-wt%: | 2.0 |
| Fixed carbon-wt%: | 25.8 |
| Ash-wt%: | 0.7 |

# Coconut shell
# (*Cocos nucifera*)

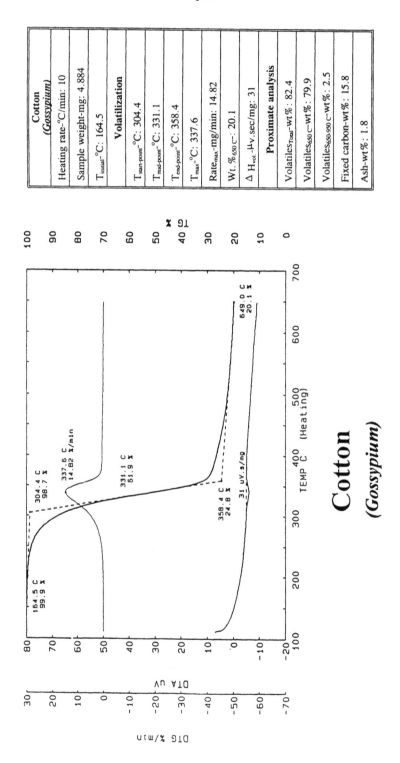

| Cotton *(Gossypium)* | |
| --- | --- |
| Heating rate-°C/min: 10 | |
| Sample weight-mg: 4.884 | |
| $T_{initial}$-°C: 164.5 | |
| **Volatilization** | |
| $T_{start-point}$-°C: 304.4 | |
| $T_{mid-point}$-°C: 331.1 | |
| $T_{end-point}$-°C: 358.4 | |
| $T_{max}$-°C: 337.6 | |
| $Rate_{max}$-mg/min: 14.82 | |
| Wt. %$_{650\,C}$: 20.1 | |
| $\Delta H_{vol}$-$\mu$v.sec/mg: 31 | |
| **Proximate analysis** | |
| Volatiles$_{Total}$-wt%: 82.4 | |
| Volatiles$_{650\,C}$-wt%: 79.9 | |
| Volatiles$_{650-950\,C}$-wt%: 2.5 | |
| Fixed carbon-wt%: 15.8 | |
| Ash-wt%: 1.8 | |

# Cotton
## *(Gossypium)*

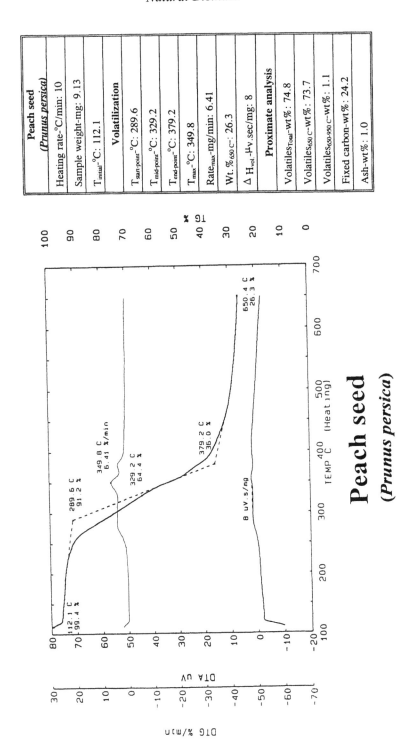

| Peach seed *(Prunus persica)* | |
|---|---|
| Heating rate-°C/min: 10 | |
| Sample weight-mg: 9.13 | |
| $T_{initial}$-°C: 112.1 | |
| **Volatilization** | |
| $T_{start-point}$-°C: 289.6 | |
| $T_{mid-point}$-°C: 329.2 | |
| $T_{end-point}$-°C: 379.2 | |
| $T_{max}$-°C: 349.8 | |
| $Rate_{max}$-mg/min: 6.41 | |
| Wt.$\%_{650°C}$: 26.3 | |
| $\Delta H_{vol.}$-$\mu$v.sec/mg: 8 | |
| **Proximate analysis** | |
| Volatiles$_{Total}$-wt%: 74.8 | |
| Volatiles$_{650°C}$-wt%: 73.7 | |
| Volatiles$_{650-950°C}$-wt%: 1.1 | |
| Fixed carbon-wt%: 24.2 | |
| Ash-wt%: 1.0 | |

# Peach seed
## *(Prunus persica)*

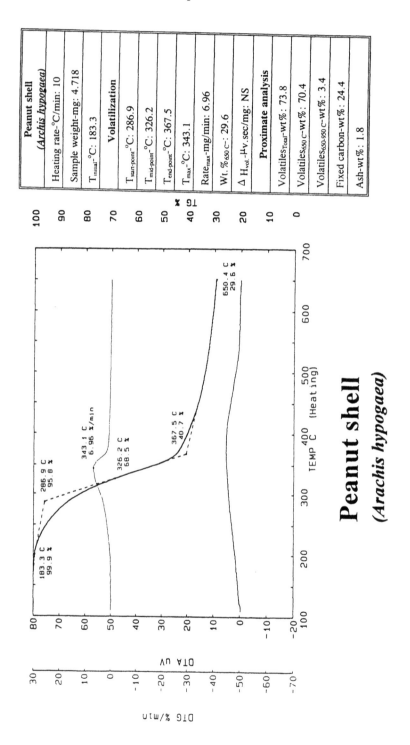

| Peanut shell *(Archis hypogaea)* | |
| --- | --- |
| Heating rate-°C/min: 10 | |
| Sample weight-mg: 4.718 | |
| $T_{initial}$-°C: 183.3 | |
| **Volatilization** | |
| $T_{start-point}$-°C: 286.9 | |
| $T_{mid-point}$-°C: 326.2 | |
| $T_{end-point}$-°C: 367.5 | |
| $T_{max}$-°C: 343.1 | |
| $Rate_{max}$-mg/min: 6.96 | |
| Wt.%$_{650 C}$: 29.6 | |
| $\Delta H_{vol.}$-$\mu$v.sec/mg: NS | |
| **Proximate analysis** | |
| Volatiles$_{Total}$-wt%: 73.8 | |
| Volatiles$_{650 C}$-wt%: 70.4 | |
| Volatiles$_{650-950 C}$-wt%: 3.4 | |
| Fixed carbon-wt%: 24.4 | |
| Ash-wt%: 1.8 | |

# Peanut shell
## *(Arachis hypogaea)*

| Pistachio nut (*Pistacia vera*) | |
|---|---|
| Heating rate-°C/min: 10 | |
| Sample weight-mg: 6.437 | |
| $T_{initial}$-°C: 192.7 | |
| **Volatilization** | |
| $T_{start-point}$-°C: 273.6 | |
| $T_{mid-point}$-°C: 319.6 | |
| $T_{end-point}$-°C: 370.0 | |
| $T_{max}$-°C: 337.7 | |
| $Rate_{max}$-mg/min: 7.34 | |
| Wt.%$_{650\,C}$: 22.9 | |
| $\Delta H_{vol}$ -μv.sec/mg: NS | |
| **Proximate analysis** | |
| Volatiles$_{Total}$-wt%: 82.3 | |
| Volatiles$_{650\,C}$-wt%: 77.1 | |
| Volatiles$_{650-950\,C}$-wt%: 5.2 | |
| Fixed carbon-wt%: 17.5 | |
| Ash-wt%: 0.2 | |

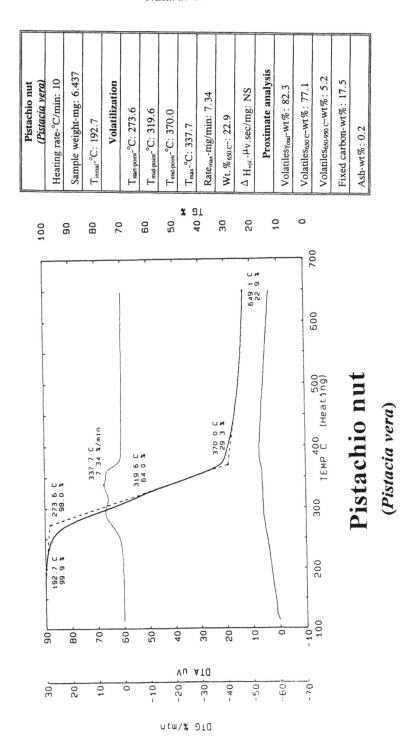

# Pistachio nut
## (*Pistacia vera*)

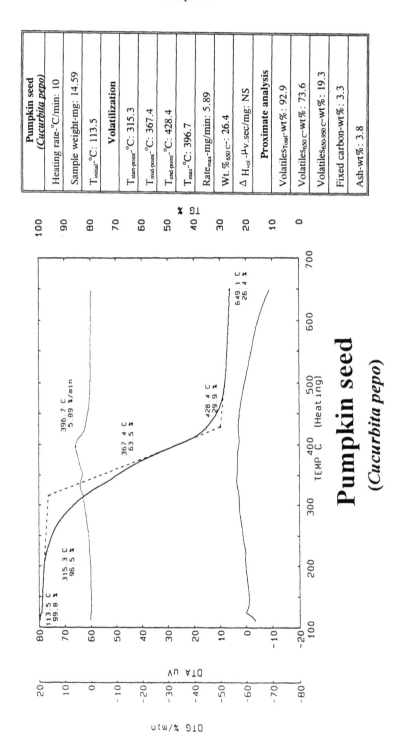

| Pumpkin seed (*Cucurbita pepo*) | |
|---|---|
| Heating rate-°C/min: | 10 |
| Sample weight-mg: | 14.59 |
| $T_{initial}$-°C: | 113.5 |
| **Volatilization** | |
| $T_{start-point}$-°C: | 315.3 |
| $T_{mid-point}$-°C: | 367.4 |
| $T_{end-point}$-°C: | 428.4 |
| $T_{max}$-°C: | 396.7 |
| $Rate_{max}$-mg/min: | 5.89 |
| Wt.%$_{650 C}$: | 26.4 |
| $\Delta H_{vol}$-$\mu$v.sec/mg: | NS |
| **Proximate analysis** | |
| Volatiles$_{Total}$-wt%: | 92.9 |
| Volatiles$_{650 C}$-wt%: | 73.6 |
| Volatiles$_{650-950 C}$-wt%: | 19.3 |
| Fixed carbon-wt%: | 3.3 |
| Ash-wt%: | 3.8 |

**Pumpkin seed**
**(*Cucurbita pepo*)**

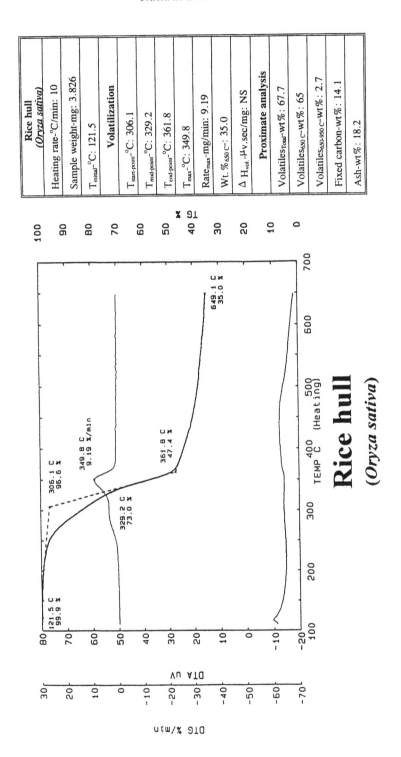

| Rice hull (*Oryza sativa*) | |
|---|---|
| Heating rate-°C/min: 10 | |
| Sample weight-mg: 3.826 | |
| $T_{initial}$-°C: 121.5 | |
| **Volatilization** | |
| $T_{start-point}$-°C: 306.1 | |
| $T_{mid-point}$-°C: 329.2 | |
| $T_{end-point}$-°C: 361.8 | |
| $T_{max}$-°C: 349.8 | |
| $Rate_{max}$-mg/min: 9.19 | |
| Wt.%$_{650°C}$: 35.0 | |
| $\Delta H_{vol}$-$\mu$v.sec/mg: NS | |
| **Proximate analysis** | |
| Volatiles$_{Total}$-wt%: 67.7 | |
| Volatiles$_{650°C}$-wt%: 65 | |
| Volatiles$_{650-950°C}$-wt%: 2.7 | |
| Fixed carbon-wt%: 14.1 | |
| Ash-wt%: 18.2 | |

# Rice hull
## (*Oryza sativa*)

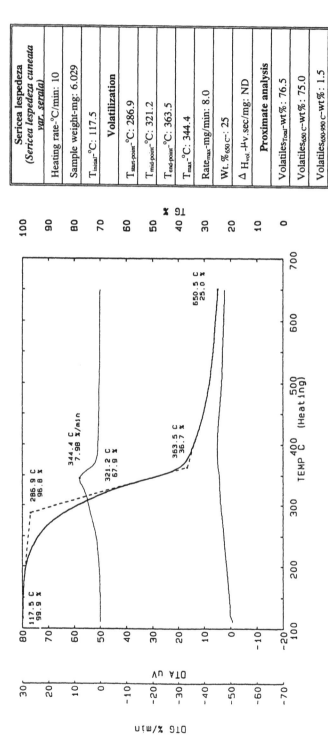

| Sericea lespedeza (Sericea lespedeza cuneata var. serala) | |
|---|---|
| Heating rate-°C/min: 10 | |
| Sample weight-mg: 6.029 | |
| $T_{initial}$-°C: 117.5 | |
| **Volatilization** | |
| $T_{start-point}$-°C: 286.9 | |
| $T_{mid-point}$-°C: 321.2 | |
| $T_{end-point}$-°C: 363.5 | |
| $T_{max}$-°C: 344.4 | |
| $Rate_{max}$-mg/min: 8.0 | |
| Wt.%$_{650 C}$: 25 | |
| $\Delta H_{vol}$-$\mu$v.sec/mg: ND | |
| **Proximate analysis** | |
| Volatiles$_{Total}$-wt%: 76.5 | |
| Volatiles$_{650 C}$-wt%: 75.0 | |
| Volatiles$_{650-950 C}$-wt%: 1.5 | |
| Fixed carbon-wt%: 21.6 | |
| Ash-wt%: 1.9 | |

# Sericea lespedeza
## (*Sericea lespedeza cuneata var. serala*)

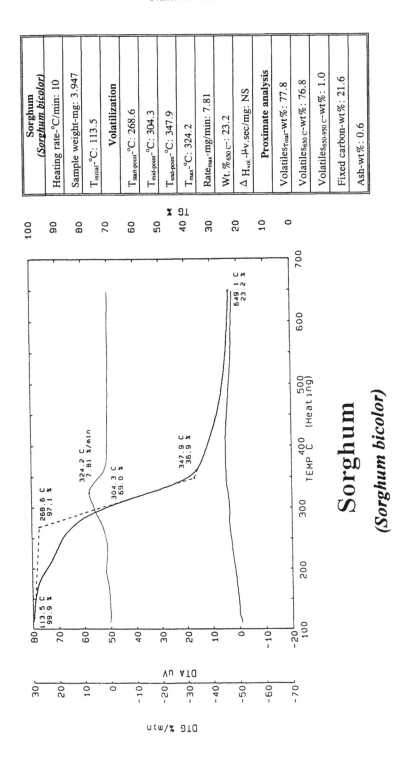

| Sorghum (*Sorghum bicolor*) | |
| --- | --- |
| Heating rate-°C/min: | 10 |
| Sample weight-mg: | 3.947 |
| $T_{initial}$-°C: | 113.5 |
| **Volatilization** | |
| $T_{start-point}$-°C: | 268.6 |
| $T_{mid-point}$-°C: | 304.3 |
| $T_{end-point}$-°C: | 347.9 |
| $T_{max}$-°C: | 324.2 |
| $Rate_{max}$-mg/min: | 7.81 |
| Wt.%$_{650 C}$: | 23.2 |
| $\Delta H_{vol.}$-$\mu$v.sec/mg: | NS |
| **Proximate analysis** | |
| Volatiles$_{Total}$-wt%: | 77.8 |
| Volatiles$_{650 C}$-wt%: | 76.8 |
| Volatiles$_{650-950 C}$-wt%: | 1.0 |
| Fixed carbon-wt%: | 21.6 |
| Ash-wt%: | 0.6 |

# Sorghum
## (*Sorghum bicolor*)

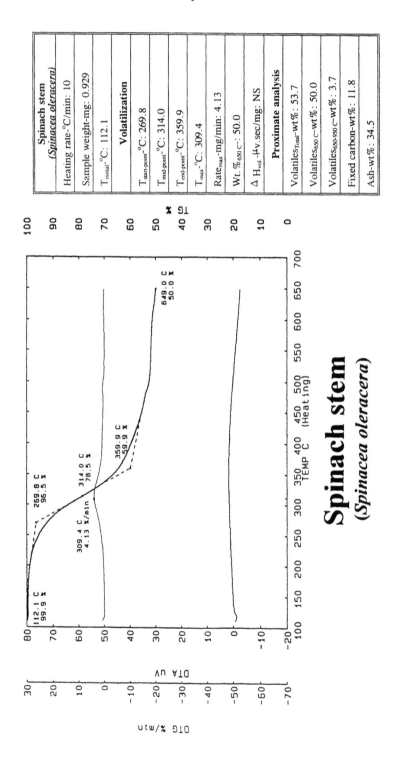

| Spinach stem<br>*(Spinacea oleracera)* | |
|---|---|
| Heating rate-°C/min: | 10 |
| Sample weight-mg: | 0.929 |
| $T_{initial}$-°C: | 112.1 |
| **Volatilization** | |
| $T_{start-point}$-°C: | 269.8 |
| $T_{mid-point}$-°C: | 314.0 |
| $T_{end-point}$-°C: | 359.9 |
| $T_{max}$-°C: | 309.4 |
| $Rate_{max}$-mg/min: | 4.13 |
| Wt.%$_{650 C}$: | 50.0 |
| $\Delta H_{vol}$-μv.sec/mg: | NS |
| **Proximate analysis** | |
| Volatiles$_{Total}$-wt%: | 53.7 |
| Volatiles$_{650 C}$-wt%: | 50.0 |
| Volatiles$_{650-950 C}$-wt%: | 3.7 |
| Fixed carbon-wt%: | 11.8 |
| Ash-wt%: | 34.5 |

# Spinach stem
## *(Spinacea oleracera)*

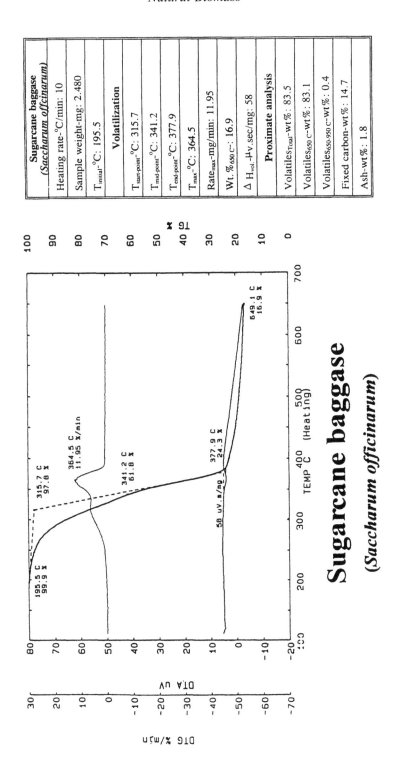

| Sugarcane baggase (*Saccharum offcinarum*) | |
|---|---|
| Heating rate-°C/min: 10 | |
| Sample weight-mg: 2.480 | |
| $T_{initial}$-°C: 195.5 | |
| **Volatilization** | |
| $T_{start-point}$-°C: 315.7 | |
| $T_{mid-point}$-°C: 341.2 | |
| $T_{end-point}$-°C: 377.9 | |
| $T_{max}$-°C: 364.5 | |
| $Rate_{max}$-mg/min: 11.95 | |
| Wt.%$_{650 C}$: 16.9 | |
| $\Delta H_{vol.}$-$\mu$v.sec/mg: 58 | |
| **Proximate analysis** | |
| Volatiles$_{Total}$-wt%: 83.5 | |
| Volatiles$_{650 C}$-wt%: 83.1 | |
| Volatiles$_{650-950 C}$-wt%: 0.4 | |
| Fixed carbon-wt%: 14.7 | |
| Ash-wt%: 1.8 | |

# Sugarcane baggase
## (*Saccharum officinarum*)

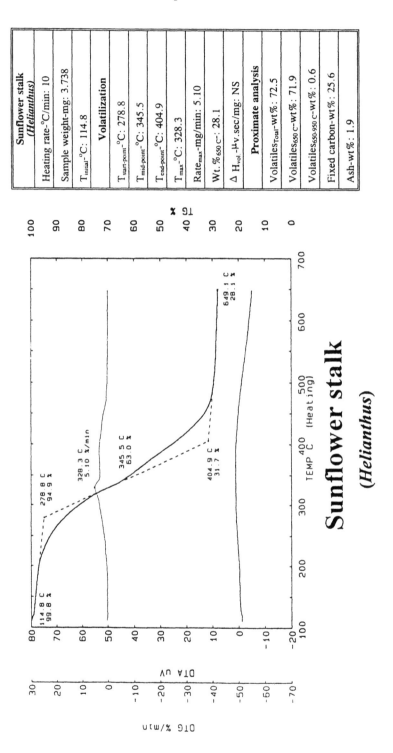

| Sunflower stalk *(Helianthus)* | |
|---|---|
| Heating rate-°C/min: 10 | |
| Sample weight-mg: 3.738 | |
| $T_{initial}$-°C: 114.8 | |
| **Volatilization** | |
| $T_{start-point}$-°C: 278.8 | |
| $T_{mid-point}$-°C: 345.5 | |
| $T_{end-point}$-°C: 404.9 | |
| $T_{max}$-°C: 328.3 | |
| $Rate_{max}$-mg/min: 5.10 | |
| Wt.%$_{650°C}$: 28.1 | |
| $\Delta H_{vol.}$-$\mu$v.sec/mg: NS | |
| **Proximate analysis** | |
| Volatiles$_{Total}$-wt%: 72.5 | |
| Volatiles$_{650°C}$-wt%: 71.9 | |
| Volatiles$_{650-950°C}$-wt%: 0.6 | |
| Fixed carbon-wt%: 25.6 | |
| Ash-wt%: 1.9 | |

# Sunflower stalk
## *(Helianthus)*

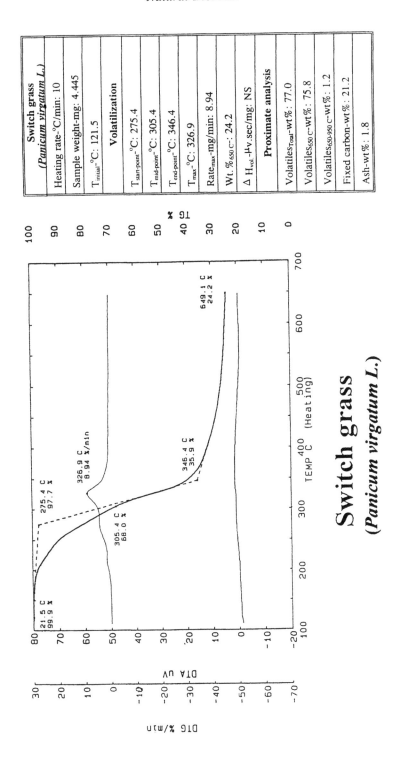

| Switch grass (Panicum virgatum L.) | |
| --- | --- |
| Heating rate-°C/min: 10 | |
| Sample weight-mg: 4.445 | |
| $T_{initial}$-°C: 121.5 | |
| **Volatilization** | |
| $T_{start-point}$-°C: 275.4 | |
| $T_{mid-point}$-°C: 305.4 | |
| $T_{end-point}$-°C: 346.4 | |
| $T_{max}$-°C: 326.9 | |
| $Rate_{max}$-mg/min: 8.94 | |
| Wt.%$_{650°C}$: 24.2 | |
| $\Delta H_{vol}$-$\mu$v.sec/mg: NS | |
| **Proximate analysis** | |
| Volatiles$_{Total}$-wt%: 77.0 | |
| Volatiles$_{650°C}$-wt%: 75.8 | |
| Volatiles$_{650-950°C}$-wt%: 1.2 | |
| Fixed carbon-wt%: 21.2 | |
| Ash-wt%: 1.8 | |

# Switch grass
## (*Panicum virgatum L.*)

*145*

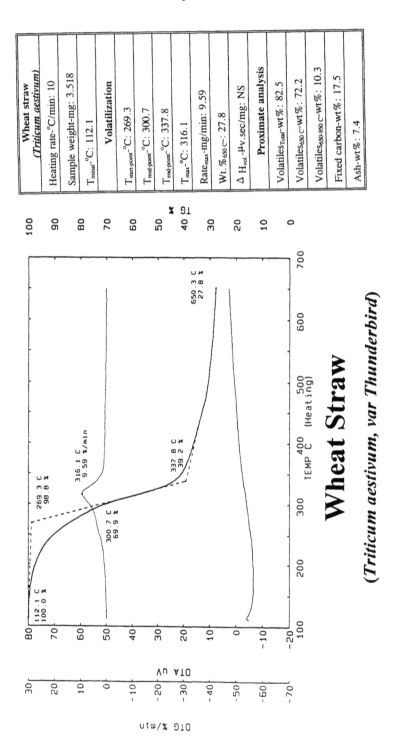

**Wheat Straw**
*(Triticum aestivum, var Thunderbird)*

| Wheat straw *(Triticum aestivum)* | |
|---|---|
| Heating rate-°C/min: | 10 |
| Sample weight-mg: | 3.518 |
| $T_{initial}$-°C: | 112.1 |
| **Volatilization** | |
| $T_{start-point}$-°C: | 269.3 |
| $T_{mid-point}$-°C: | 300.7 |
| $T_{end-point}$-°C: | 337.8 |
| $T_{max}$-°C: | 316.1 |
| $Rate_{max}$-mg/min: | 9.59 |
| Wt.%$_{650C}$: | 27.8 |
| $\Delta H_{vol}$-$\mu$v.sec/mg: | NS |
| **Proximate analysis** | |
| Volatiles$_{Total}$-wt%: | 82.5 |
| Volatiles$_{650C}$-wt%: | 72.2 |
| Volatiles$_{650-950C}$-wt%: | 10.3 |
| Fixed carbon-wt%: | 17.5 |
| Ash-wt%: | 7.4 |

# AQUATIC BIOMASS

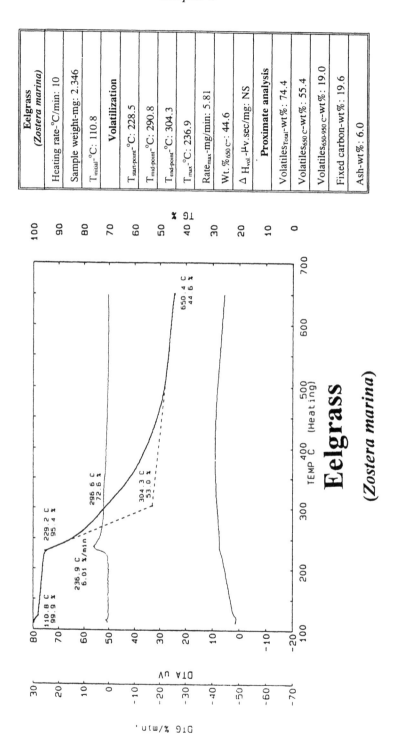

| Eelgrass (Zostera marina) | |
|---|---|
| Heating rate-°C/min: 10 | |
| Sample weight-mg: 2.346 | |
| $T_{initial}$-°C: 110.8 | |
| **Volatilization** | |
| $T_{start-point}$-°C: 228.5 | |
| $T_{mid-point}$-°C: 290.8 | |
| $T_{end-point}$-°C: 304.3 | |
| $T_{max}$-°C: 236.9 | |
| $Rate_{max}$-mg/min: 5.81 | |
| $Wt.\%_{650\,C}$: 44.6 | |
| $\Delta H_{vol}$-$\mu$v.sec/mg: NS | |
| **Proximate analysis** | |
| $Volatiles_{Total}$-wt%: 74.4 | |
| $Volatiles_{650\,C}$-wt%: 55.4 | |
| $Volatiles_{650-950\,C}$-wt%: 19.0 | |
| Fixed carbon-wt%: 19.6 | |
| Ash-wt%: 6.0 | |

# Eelgrass
## (Zostera marina)

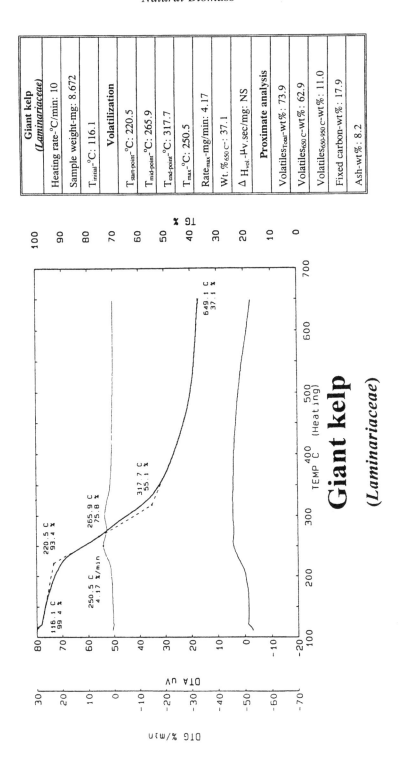

| Giant kelp (*Laminariaceae*) | |
|---|---|
| Heating rate-°C/min: | 10 |
| Sample weight-mg: | 8.672 |
| $T_{initial}$-°C: | 116.1 |
| **Volatilization** | |
| $T_{start-point}$-°C: | 220.5 |
| $T_{mid-point}$-°C: | 265.9 |
| $T_{end-point}$-°C: | 317.7 |
| $T_{max}$-°C: | 250.5 |
| $Rate_{max}$-mg/min: | 4.17 |
| Wt.%$_{650°C}$: | 37.1 |
| $\Delta H_{vol}$-$\mu$v.sec/mg: | NS |
| **Proximate analysis** | |
| Volatiles$_{T_{coal}}$-wt%: | 73.9 |
| Volatiles$_{650°C}$-wt%: | 62.9 |
| Volatiles$_{650-950°C}$-wt%: | 11.0 |
| Fixed carbon-wt%: | 17.9 |
| Ash-wt%: | 8.2 |

# Giant kelp
## (*Laminariaceae*)

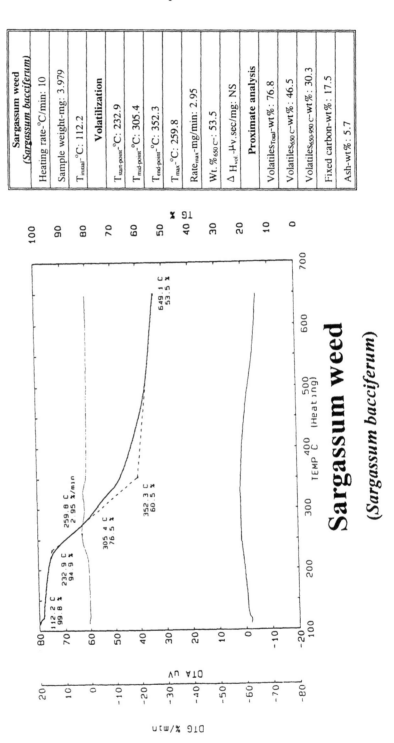

| Sargassum weed (*Sargassum bacciferum*) | |
|---|---|
| Heating rate-°C/min: 10 | |
| Sample weight-mg: 3.979 | |
| $T_{initial}$-°C: 112.2 | |
| **Volatilization** | |
| $T_{start-point}$-°C: 232.9 | |
| $T_{mid-point}$-°C: 305.4 | |
| $T_{end-point}$-°C: 352.3 | |
| $T_{max}$-°C: 259.8 | |
| $Rate_{max}$-mg/min: 2.95 | |
| Wt. %$_{650 C}$: 53.5 | |
| $\Delta H_{vol}$-μv.sec/mg: NS | |
| **Proximate analysis** | |
| Volatiles$_{Total}$-wt%: 76.8 | |
| Volatiles$_{650 C}$-wt%: 46.5 | |
| Volatiles$_{650-950 C}$-wt%: 30.3 | |
| Fixed carbon-wt%: 17.5 | |
| Ash-wt%: 5.7 | |

## Sargassum weed
### (*Sargassum bacciferum*)

# 7

# VARIATIONS IN BIOMASS

## 7.1 EFFECT OF PLANT STORAGE

### 7.1.1 Storage Under Controlled Conditions

A number of species listed in this chapter were supplied by the Natural Renewable Energy Laboratory (NREL) of Golden, CO, to test for the effect of storage on biomass fuels. These samples were stored at NREL under controlled conditions where fresh samples were at first divided into two parts: One part was refrigerated and the other was stored at room temperature for half a year (26 weeks) to see the effect of storage.

Examination of the several thermograms presented in this chapter shows that storage increased the volatility and decreased the fixed carbon content in most of the samples (except for bagasse). In addition, a substantial increase in the ash content was observed in all of the samples that were stored under room temperature conditions.

### 7.1.2 Storage Under Uncontrolled Conditions

When wood is left to decompose in the forest or in any other uncontrolled conditions, typically the cellulose and hemicellulose are decomposed, leaving a form of lignin thought to be relatively unaltered. The thermograms of brown and white rotten woods represent the thermal behavior for these type of samples. The two thermograms are very similar to the other lignins shown in Chapter 5 but have a much higher ash content.

## 7.2   EFFECT OF PLANT ANATOMY

The various parts of plants have different compositions appropriate to their structures and functions, as discussed in Chapters 1 and 6. In this chapter, we show the differences in thermal degradation pattern for various structures present in a spinach plant.

## 7.3   EFFECT OF GROWTH CONDITIONS

In several studies, it is suggested that the conditions of growth affects plant composition in terms of volatiles, ash content, and fixed carbon. In this chapter, we show the affect of irrigation on the thermal behavior of cotton. The two thermograms are very similar to each other and do not show any significant change that can be attributed to this factor. However, we believe more samples need to be examined under this category.

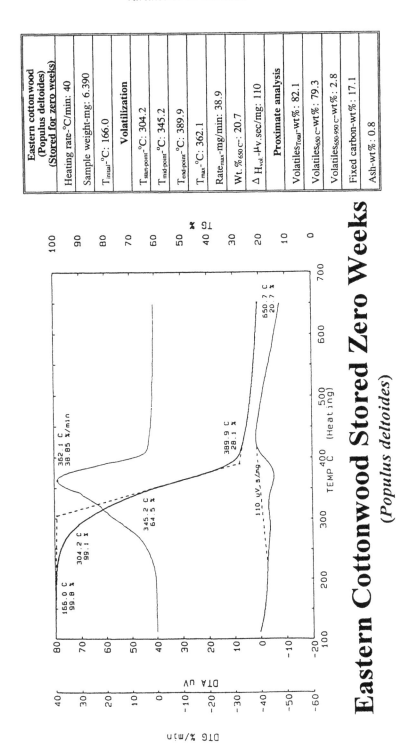

| Eastern cottonwood (Populus deltoides) (Stored for zero weeks) | |
| --- | --- |
| Heating rate-°C/min: 40 | |
| Sample weight-mg: 6.390 | |
| $T_{initial}$-°C: 166.0 | |
| **Volatilization** | |
| $T_{start-point}$-°C: 304.2 | |
| $T_{mid-point}$-°C: 345.2 | |
| $T_{end-point}$-°C: 389.9 | |
| $T_{max}$-°C: 362.1 | |
| $Rate_{max}$-mg/min: 38.9 | |
| Wt.%$_{650 C}$: 20.7 | |
| $\Delta H_{vol}$-µv.sec/mg: 110 | |
| **Proximate analysis** | |
| Volatiles$_{Total}$-wt%: 82.1 | |
| Volatiles$_{650 C}$-wt%: 79.3 | |
| Volatiles$_{650-950 C}$-wt%: 2.8 | |
| Fixed carbon-wt%: 17.1 | |
| Ash-wt%: 0.8 | |

## Eastern Cottonwood Stored Zero Weeks
### (*Populus deltoides*)

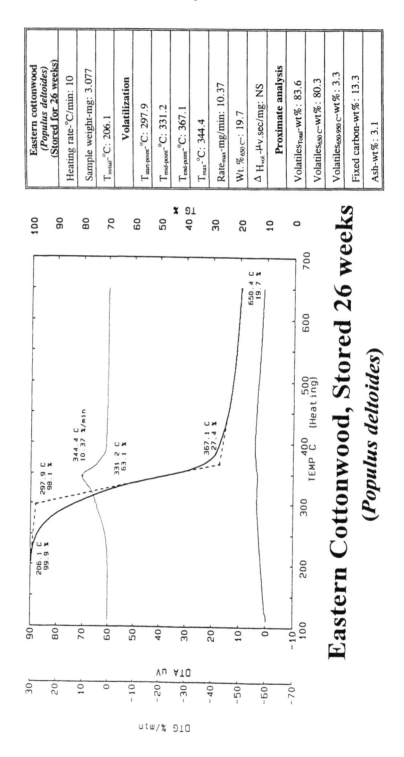

| Eastern cottonwood *(Populus deltoides)* (Stored for 26 weeks) | |
|---|---|
| Heating rate-°C/min: 10 | |
| Sample weight-mg: 3.077 | |
| $T_{initial}$-°C: 206.1 | |
| **Volatilization** | |
| $T_{start-point}$-°C: 297.9 | |
| $T_{mid-point}$-°C: 331.2 | |
| $T_{end-point}$-°C: 367.1 | |
| $T_{max}$-°C: 344.4 | |
| $Rate_{max}$-mg/min: 10.37 | |
| Wt.%$_{650\,c}$: 19.7 | |
| $\Delta\,H_{vol}$-Hv.sec/mg: NS | |
| **Proximate analysis** | |
| Volatiles$_{Total}$-wt%: 83.6 | |
| Volatiles$_{650\,c}$-wt%: 80.3 | |
| Volatiles$_{650-950\,c}$-wt%: 3.3 | |
| Fixed carbon-wt%: 13.3 | |
| Ash-wt%: 3.1 | |

# Eastern Cottonwood, Stored 26 weeks
## *(Populus deltoides)*

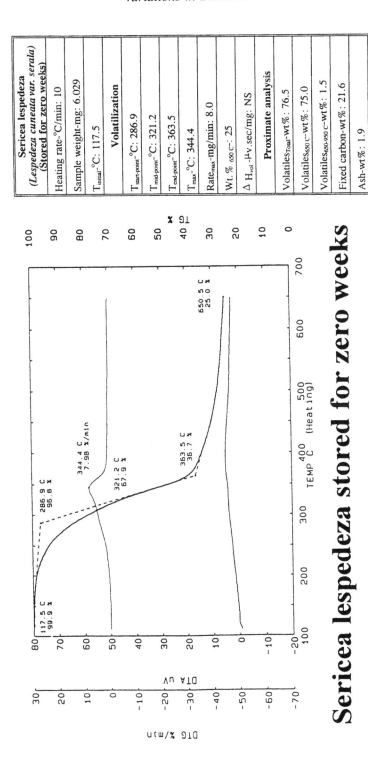

| Sericea lespedeza |
| *(Lespedeza cuneata var. serala)* |
| (Stored for zero weeks) |
| Heating rate-°C/min: 10 |
| Sample weight-mg: 6.029 |
| $T_{initial}$-°C: 117.5 |
| **Volatilization** |
| $T_{start-point}$-°C: 286.9 |
| $T_{mid-point}$-°C: 321.2 |
| $T_{end-point}$-°C: 363.5 |
| $T_{max}$-°C: 344.4 |
| $Rate_{max}$-mg/min: 8.0 |
| Wt. % $_{650°C}$: 25 |
| $\Delta H_{vol}$-µv.sec/mg: NS |
| **Proximate analysis** |
| Volatiles$_{Total}$-wt%: 76.5 |
| Volatiles$_{650°C}$-wt%: 75.0 |
| Volatiles$_{650-950°C}$-wt%: 1.5 |
| Fixed carbon-wt%: 21.6 |
| Ash-wt%: 1.9 |

# Sericea lespedeza stored for zero weeks

## *(Lespedeza cuneata var. serala)*

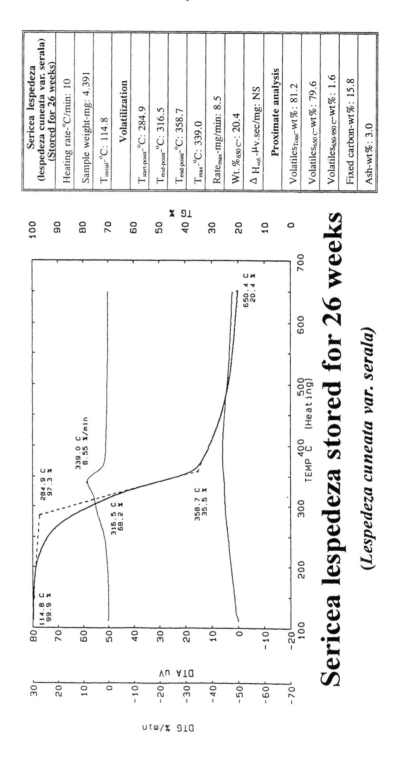

| Sericea lespedeza (lespedeza cuneata var. serala) (Stored for 26 weeks) | |
| --- | --- |
| Heating rate-°C/min: 10 | |
| Sample weight-mg: 4.391 | |
| $T_{initial}$-°C: 114.8 | |
| **Volatilization** | |
| $T_{start-point}$-°C: 284.9 | |
| $T_{mid-point}$-°C: 316.5 | |
| $T_{end-point}$-°C: 358.7 | |
| $T_{max}$-°C: 339.0 | |
| $Rate_{max}$-mg/min: 8.5 | |
| Wt.%$_{650°C}$: 20.4 | |
| $\Delta H_{vol.}$-$\mu$v.sec/mg: NS | |
| **Proximate analysis** | |
| Volatiles$_{Total}$-wt%: 81.2 | |
| Volatiles$_{650°C}$-wt%: 79.6 | |
| Volatiles$_{650-950°C}$-wt%: 1.6 | |
| Fixed carbon-wt%: 15.8 | |
| Ash-wt%: 3.0 | |

# Sericea lespedeza stored for 26 weeks
## *(Lespedeza cuneata var. serala)*

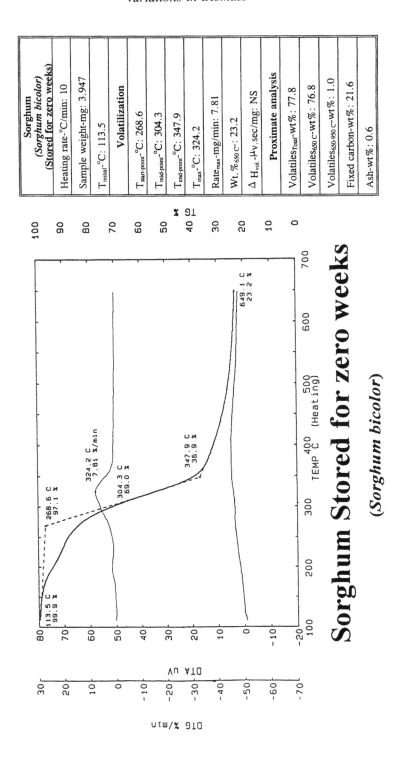

| Sorghum<br>*(Sorghum bicolor)*<br>(Stored for zero weeks) | |
|---|---|
| Heating rate-°C/min: 10 | |
| Sample weight-mg: 3.947 | |
| $T_{initial}$-°C: 113.5 | |
| **Volatilization** | |
| $T_{start-point}$-°C: 268.6 | |
| $T_{mid-point}$-°C: 304.3 | |
| $T_{end-point}$-°C: 347.9 | |
| $T_{max}$-°C: 324.2 | |
| $Rate_{max}$-mg/min: 7.81 | |
| $Wt.\%_{650 C}$: 23.2 | |
| $\Delta H_{vol}$-$\mu$v.sec/mg: NS | |
| **Proximate analysis** | |
| $Volatiles_{Total}$-wt%: 77.8 | |
| $Volatiles_{650 C}$-wt%: 76.8 | |
| $Volatiles_{650-950 C}$-wt%: 1.0 | |
| Fixed carbon-wt%: 21.6 | |
| Ash-wt%: 0.6 | |

## Sorghum Stored for zero weeks

### *(Sorghum bicolor)*

157

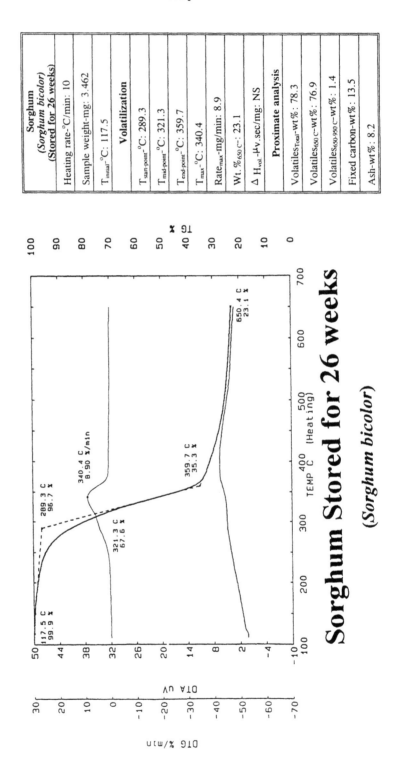

| Sorghum (*Sorghum bicolor*) (Stored for 26 weeks) | |
| --- | --- |
| Heating rate-°C/min: | 10 |
| Sample weight-mg: | 3.462 |
| $T_{initial}$-°C: | 117.5 |
| **Volatilization** | |
| $T_{start-point}$-°C: | 289.3 |
| $T_{mid-point}$-°C: | 321.3 |
| $T_{end-point}$-°C: | 359.7 |
| $T_{max}$-°C: | 340.4 |
| $Rate_{max}$-mg/min: | 8.9 |
| Wt.%$_{650}$°C: | 23.1 |
| $\Delta H_{vol}$-$\mu$v.sec/mg: | NS |
| **Proximate analysis** | |
| Volatiles$_{Total}$-wt%: | 78.3 |
| Volatiles$_{650}$°C-wt%: | 76.9 |
| Volatiles$_{650-950}$°C-wt%: | 1.4 |
| Fixed carbon-wt%: | 13.5 |
| Ash-wt%: | 8.2 |

# Sorghum Stored for 26 weeks
## (*Sorghum bicolor*)

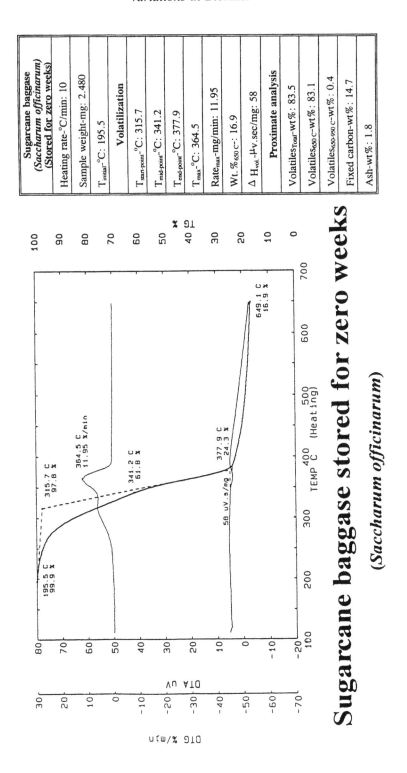

| Sugarcane baggase (*Saccharum officinarum*) (Stored for zero weeks) | |
|---|---|
| Heating rate-°C/min: 10 | |
| Sample weight-mg: 2.480 | |
| $T_{initial}$-°C: 195.5 | |
| **Volatilization** | |
| $T_{start-point}$-°C: 315.7 | |
| $T_{mid-point}$-°C: 341.2 | |
| $T_{end-point}$-°C: 377.9 | |
| $T_{max}$-°C: 364.5 | |
| $Rate_{max}$-mg/min: 11.95 | |
| Wt.%$_{650 C}$: 16.9 | |
| $\Delta H_{vol}$-$\mu$v.sec/mg: 58 | |
| **Proximate analysis** | |
| Volatiles$_{Total}$-wt%: 83.5 | |
| Volatiles$_{650 C}$-wt%: 83.1 | |
| Volatiles$_{650-950 C}$-wt%: 0.4 | |
| Fixed carbon-wt%: 14.7 | |
| Ash-wt%: 1.8 | |

## Sugarcane baggase stored for zero weeks

### (*Saccharum officinarum*)

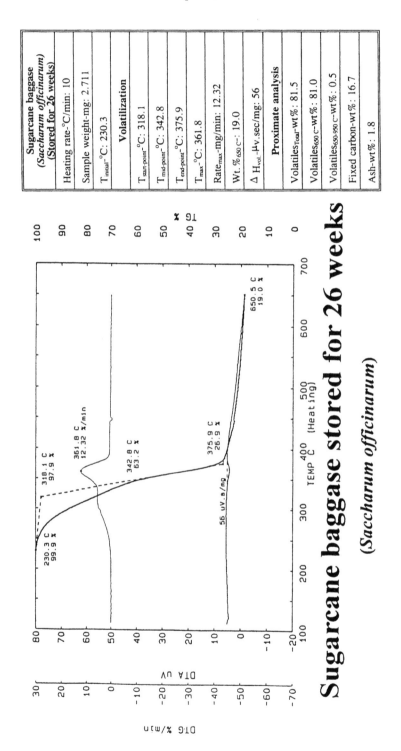

| Sugarcane baggase (*Saccharum officinarum*) (Stored for 26 weeks) |
| --- |
| Heating rate-°C/min: 10 |
| Sample weight-mg: 2.711 |
| $T_{initial}$-°C: 230.3 |
| **Volatilization** |
| $T_{start-point}$-°C: 318.1 |
| $T_{mid-point}$-°C: 342.8 |
| $T_{end-point}$-°C: 375.9 |
| $T_{max}$-°C: 361.8 |
| $Rate_{max}$-mg/min: 12.32 |
| Wt.%$_{650 C}$: 19.0 |
| $\Delta H_{vol.}$-$\mu$v. sec/mg: 56 |
| **Proximate analysis** |
| Volatiles$_{Total}$-wt%: 81.5 |
| Volatiles$_{650 C}$-wt%: 81.0 |
| Volatiles$_{650-950 C}$-wt%: 0.5 |
| Fixed carbon-wt%: 16.7 |
| Ash-wt%: 1.8 |

# Sugarcane baggase stored for 26 weeks
## (*Saccharum officinarum*)

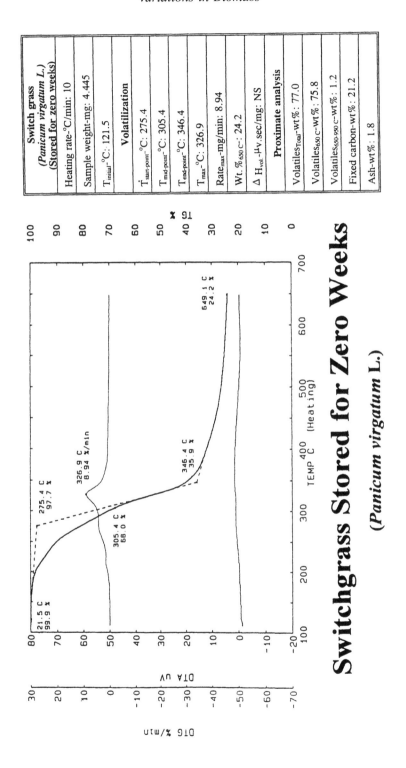

| Switch grass *(Panicum virgatum L.)* (Stored for zero weeks) | |
|---|---|
| Heating rate-°C/min: | 10 |
| Sample weight-mg: | 4.445 |
| $T_{initial}$-°C: | 121.5 |
| **Volatilization** | |
| $T_{start-point}$-°C: | 275.4 |
| $T_{mid-point}$-°C: | 305.4 |
| $T_{end-point}$-°C: | 346.4 |
| $T_{max}$-°C: | 326.9 |
| Rate$_{max}$-mg/min: | 8.94 |
| Wt.%$_{650C}$: | 24.2 |
| $\Delta H_{vol}$-$\mu$v.sec/mg: | NS |
| **Proximate analysis** | |
| Volatiles$_{Total}$-wt%: | 77.0 |
| Volatiles$_{650C}$-wt%: | 75.8 |
| Volatiles$_{650-950C}$-wt%: | 1.2 |
| Fixed carbon-wt%: | 21.2 |
| Ash-wt%: | 1.8 |

# Switchgrass Stored for Zero Weeks
## *(Panicum virgatum L.)*

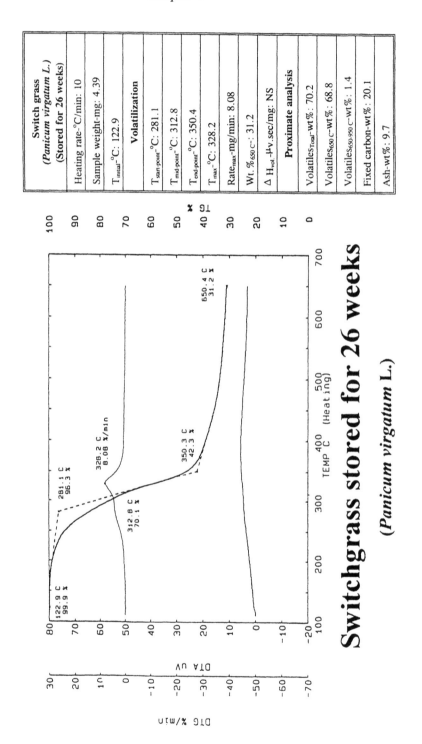

| Switch grass (*Panicum virgatum L.*) (Stored for 26 weeks) |
| --- |
| Heating rate-°C/min: 10 |
| Sample weight-mg: 4.39 |
| $T_{initial}$-°C: 122.9 |
| **Volatilization** |
| $T_{start-point}$-°C: 281.1 |
| $T_{mid-point}$-°C: 312.8 |
| $T_{end-point}$-°C: 350.4 |
| $T_{max}$-°C: 328.2 |
| $Rate_{max}$-mg/min: 8.08 |
| Wt.$\%_{650°C}$: 31.2 |
| $\Delta H_{vol}$-$\mu$v.sec/mg: NS |
| **Proximate analysis** |
| $Volatiles_{Total}$-wt%: 70.2 |
| $Volatiles_{650C}$-wt%: 68.8 |
| $Volatiles_{650-950C}$-wt%: 1.4 |
| Fixed carbon-wt%: 20.1 |
| Ash-wt%: 9.7 |

# Switchgrass stored for 26 weeks
## (*Panicum virgatum L.*)

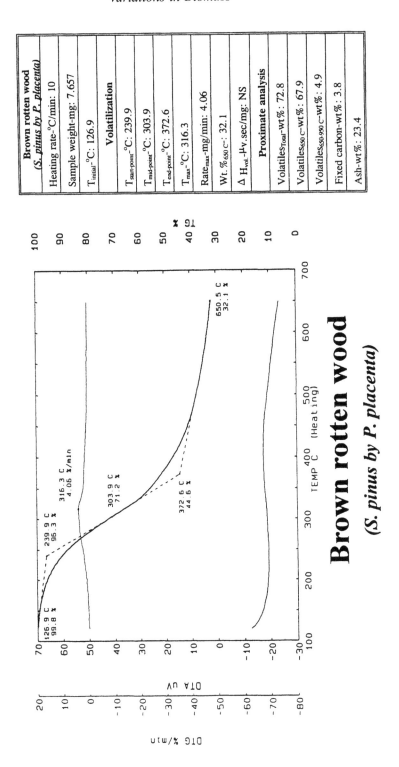

| Brown rotten wood (S. pinus by P. placenta) | |
|---|---|
| Heating rate-°C/min: | 10 |
| Sample weight-mg: | 7.657 |
| $T_{initial}$-°C: | 126.9 |
| **Volatilization** | |
| $T_{start-point}$-°C: | 239.9 |
| $T_{mid-point}$-°C: | 303.9 |
| $T_{end-point}$-°C: | 372.6 |
| $T_{max}$-°C: | 316.3 |
| $Rate_{max}$-mg/min: | 4.06 |
| Wt.%$_{650°C}$: | 32.1 |
| $\Delta H_{vol.}$-$\mu$v.sec/mg: | NS |
| **Proximate analysis** | |
| Volatiles$_{Total}$-wt%: | 72.8 |
| Volatiles$_{650°C}$-wt%: | 67.9 |
| Volatiles$_{650-950°C}$-wt%: | 4.9 |
| Fixed carbon-wt%: | 3.8 |
| Ash-wt%: | 23.4 |

**Brown rotten wood**

*(S. pinus by P. placenta)*

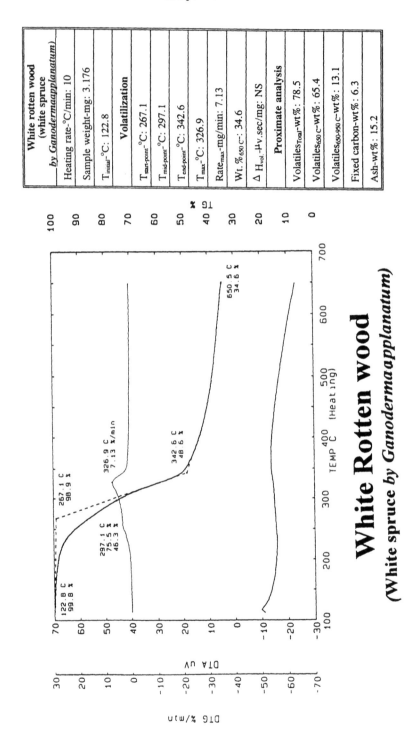

| White rotten wood (white spruce by *Ganodermaapplanatum*) | |
| --- | --- |
| Heating rate-°C/min: | 10 |
| Sample weight-mg: | 3.176 |
| $T_{initial}$-°C: | 122.8 |
| **Volatilization** | |
| $T_{start-point}$-°C: | 267.1 |
| $T_{mid-point}$-°C: | 297.1 |
| $T_{end-point}$-°C: | 342.6 |
| $T_{max}$-°C: | 326.9 |
| $Rate_{max}$-mg/min: | 7.13 |
| Wt.%$_{650 C}$: | 34.6 |
| $\Delta H_{vol.}$-$\mu$v.sec/mg: | NS |
| **Proximate analysis** | |
| Volatiles$_{Total}$-wt%: | 78.5 |
| Volatiles$_{650 C}$-wt%: | 65.4 |
| Volatiles$_{650-950 C}$-wt%: | 13.1 |
| Fixed carbon-wt%: | 6.3 |
| Ash-wt%: | 15.2 |

# White Rotten wood
## (White spruce by *Ganodermaapplanatum*)

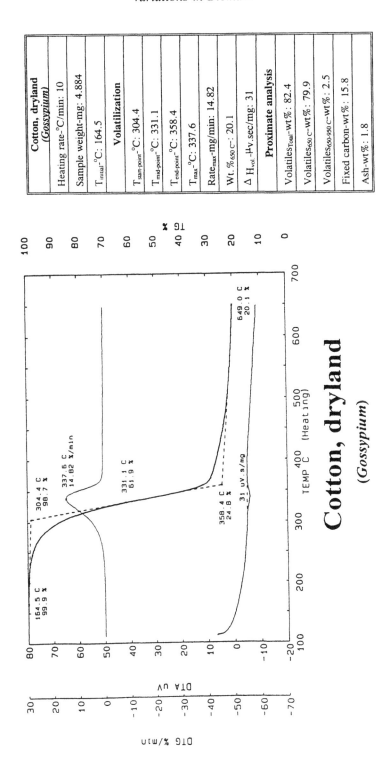

| Cotton, dryland (*Gossypium*) | |
|---|---|
| Heating rate-°C/min: 10 | |
| Sample weight-mg: 4.884 | |
| $T_{initial}$-°C: 164.5 | |
| **Volatilization** | |
| $T_{start-point}$-°C: 304.4 | |
| $T_{mid-point}$-°C: 331.1 | |
| $T_{end-point}$-°C: 358.4 | |
| $T_{max}$-°C: 337.6 | |
| $Rate_{max}$-mg/min: 14.82 | |
| Wt. %$_{650 C}$: 20.1 | |
| $\Delta H_{vol}$-$\mu$v.sec/mg: 31 | |
| **Proximate analysis** | |
| Volatiles$_{Total}$-wt%: 82.4 | |
| Volatiles$_{650 C}$-wt%: 79.9 | |
| Volatiles$_{650-950 C}$-wt%: 2.5 | |
| Fixed carbon-wt%: 15.8 | |
| Ash-wt%: 1.8 | |

# Cotton, dryland

## (*Gossypium*)

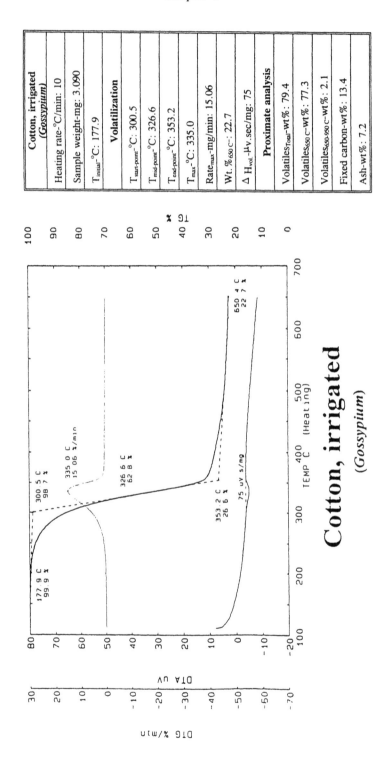

| Cotton, irrigated (*Gossypium*) |
|---|
| Heating rate-°C/min: 10 |
| Sample weight-mg: 3.090 |
| T$_{initial}$-°C: 177.9 |
| **Volatilization** |
| T$_{start-point}$-°C: 300.5 |
| T$_{mid-point}$-°C: 326.6 |
| T$_{end-point}$-°C: 353.2 |
| T$_{max}$-°C: 335.0 |
| Rate$_{max}$-mg/min: 15.06 |
| Wt.%$_{650°C}$: 22.7 |
| $\Delta$ H$_{vol}$-$\mu$v.sec/mg: 75 |
| **Proximate analysis** |
| Volatiles$_{Total}$-wt%: 79.4 |
| Volatiles$_{650°C}$-wt%: 77.3 |
| Volatiles$_{650-950°C}$-wt%: 2.1 |
| Fixed carbon-wt%: 13.4 |
| Ash-wt%: 7.2 |

**Cotton, irrigated**
(*Gossypium*)

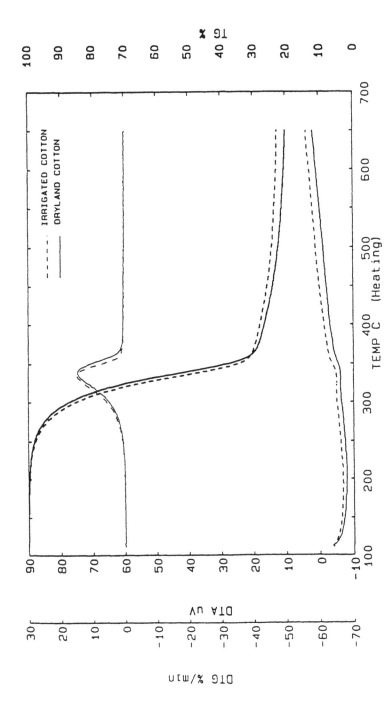

## Effect of growth conditions
### (Dryland and Irrigated Cotton)

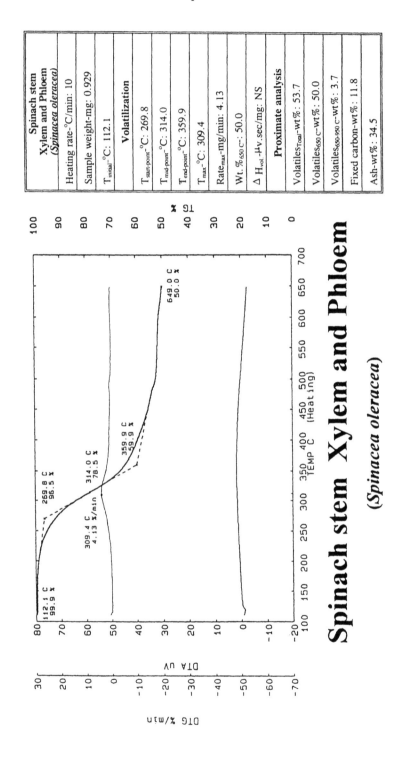

# Spinach stem Xylem and Phloem
## (*Spinacea oleracea*)

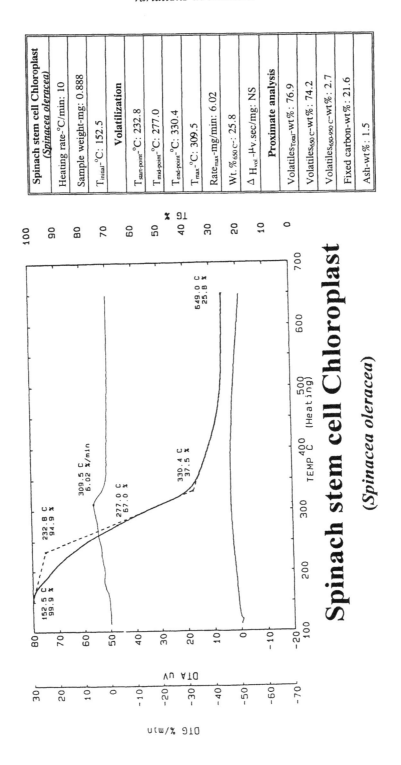

| Spinach stem cell Chloroplast<br>*(Spinacea oleracea)* |
| --- |
| Heating rate–°C/min: 10 |
| Sample weight–mg: 0.888 |
| $T_{initial}$–°C: 152.5 |
| **Volatilization** |
| $T_{start-point}$–°C: 232.8 |
| $T_{mid-point}$–°C: 277.0 |
| $T_{end-point}$–°C: 330.4 |
| $T_{max}$–°C: 309.5 |
| $Rate_{max}$–mg/min: 6.02 |
| Wt.%$_{650C}$: 25.8 |
| $\Delta H_{vol}$–$\mu$v.sec/mg: NS |
| **Proximate analysis** |
| Volatiles$_{Total}$–wt%: 76.9 |
| Volatiles$_{650C}$–wt%: 74.2 |
| Volatiles$_{650-950C}$–wt%: 2.7 |
| Fixed carbon–wt%: 21.6 |
| Ash–wt%: 1.5 |

# Spinach stem cell Chloroplast
## *(Spinacea oleracea)*

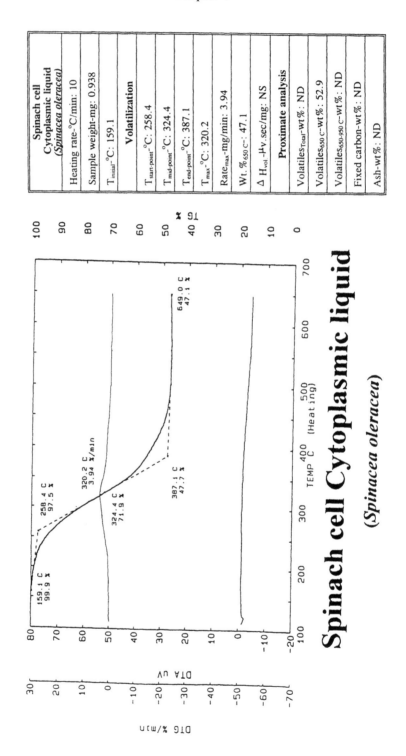

| Spinach cell Cytoplasmic liquid (*Spinacea oleracea*) | |
| --- | --- |
| Heating rate-°C/min: | 10 |
| Sample weight-mg: | 0.938 |
| $T_{initial}$-°C: | 159.1 |
| **Volatilization** | |
| $T_{start-point}$-°C: | 258.4 |
| $T_{mid-point}$-°C: | 324.4 |
| $T_{end-point}$-°C: | 387.1 |
| $T_{max}$-°C: | 320.2 |
| $Rate_{max}$-mg/min: | 3.94 |
| Wt. $\%_{650\,C}$-: | 47.1 |
| $\Delta H_{vol}$-$\mu$v.sec/mg: | NS |
| **Proximate analysis** | |
| $Volatiles_{Total}$-wt%: | ND |
| $Volatiles_{650\,C}$-wt%: | 52.9 |
| $Volatiles_{650-950\,C}$-wt%: | ND |
| Fixed carbon-wt%: | ND |
| Ash-wt%: | ND |

# Spinach cell Cytoplasmic liquid
## (*Spinacea oleracea*)

# 8

# PROCESSED BIOMASS

## 8.1  PAPER

The manufacture of paper represents the largest use of wood world-wide. Paper can also be manufactured from other biomass fibers such as rags or kenaff (Weber, 1982). In general, newsprint is relatively low in ash, while magazine grade paper is high in ash (i.e., 30%) due to the clay or other filler used for whiteness. Chemical pulping produces a business paper from softwood called lightweight paper; that produced from hardwoods is called heavy paper, for example, manila folders.

The paper samples used in this book were obtained from The Institute for Paper Science and Technology (IPST), located at the Georgia Institute of Technology in Atlanta, GA.

The thermograms of different paper samples appear to be similar to each other, as expected. Since paper is primarily composed of the cellulose fibers of wood, it is also not surprising that the thermograms of all the samples have midpoints in the 315–350°C range, like cellulose itself. The only difference in thermograms is in terms of total ash content, which is due to different additives in the processing of various types of papers.

## 8.2  DENSIFIED BIOMASS

Biomass occurs in many forms, often of very low density, that are unsuitable for storage and transportation as fuel. The density of biomass can be increased many times by briquetting, cubing, or pelletizing processes in which pressures up to $10^3$ atmospheres squeeze out the void space and cement the particles together (Reed, 1978; Bain, 1981).

A thermogram of Western red ceder is presented as an example of pelletized soft wood, while that of Osage orange is shown as an example of pelletized hard wood. The two thermograms show that the thermal decomposition parameters are generally not affected by these types of physical processing as long as the sample Biot number is kept below the recommended value of 0.1

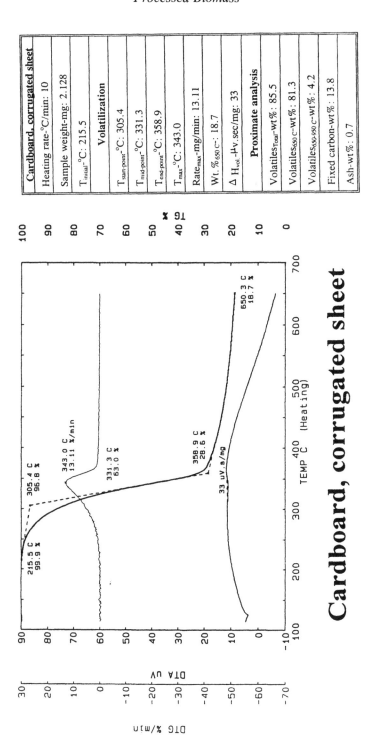

| Cardboard, corrugated sheet | |
|---|---|
| Heating rate-°C/min: 10 | |
| Sample weight-mg: 2.128 | |
| $T_{initial}$-°C: 215.5 | |
| **Volatilization** | |
| $T_{start-point}$-°C: 305.4 | |
| $T_{mid-point}$-°C: 331.3 | |
| $T_{end-point}$-°C: 358.9 | |
| $T_{max}$-°C: 343.0 | |
| $Rate_{max}$-mg/min: 13.11 | |
| $Wt.\%_{650°C}$: 18.7 | |
| $\Delta H_{vol}$.-$\mu$v.sec/mg: 33 | |
| **Proximate analysis** | |
| $Volatiles_{Total}$-wt%: 85.5 | |
| $Volatiles_{650°C}$-wt%: 81.3 | |
| $Volatiles_{650-950°C}$-wt%: 4.2 | |
| Fixed carbon-wt%: 13.8 | |
| Ash-wt%: 0.7 | |

# Cardboard, corrugated sheet

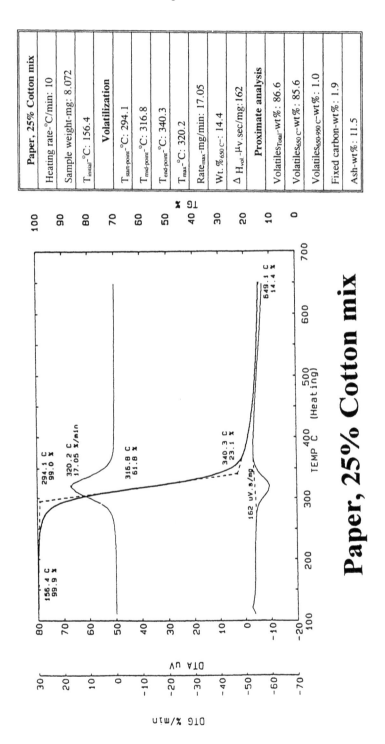

| Paper, 25% Cotton mix | |
|---|---|
| Heating rate-°C/min: 10 | |
| Sample weight-mg: 8.072 | |
| $T_{initial}$°C: 156.4 | |
| **Volatilization** | |
| $T_{start-point}$-°C: 294.1 | |
| $T_{mid-point}$-°C: 316.8 | |
| $T_{end-point}$-°C: 340.3 | |
| $T_{max}$-°C: 320.2 | |
| $Rate_{max}$-mg/min: 17.05 | |
| Wt.%$_{650 C}$: 14.4 | |
| $\Delta H_{vol}$-Hv.sec/mg:162 | |
| **Proximate analysis** | |
| Volatiles$_{Total}$-wt%: 86.6 | |
| Volatiles$_{650 C}$-wt%: 85.6 | |
| Volatiles$_{650-950 C}$-wt%: 1.0 | |
| Fixed carbon-wt%: 1.9 | |
| Ash-wt%: 11.5 | |

## Paper, 25% Cotton mix

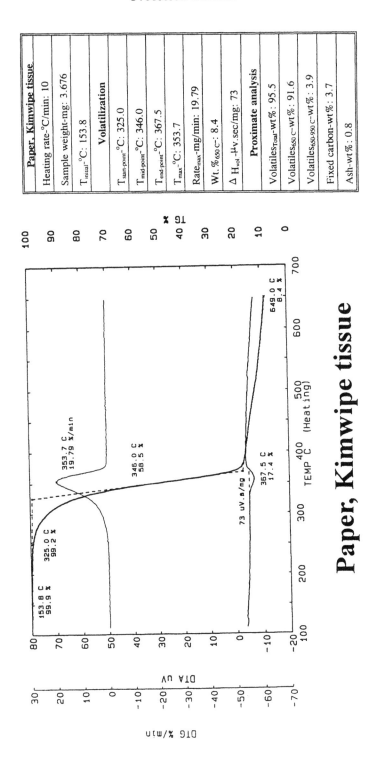

| Paper, Kimwipe tissue | |
|---|---|
| Heating rate-°C/min: 10 | |
| Sample weight-mg: 3.676 | |
| $T_{initial}$-°C: 153.8 | |
| **Volatilization** | |
| $T_{start-point}$-°C: 325.0 | |
| $T_{mid-point}$-°C: 346.0 | |
| $T_{end-point}$-°C: 367.5 | |
| $T_{max}$-°C: 353.7 | |
| $Rate_{max}$-mg/min: 19.79 | |
| Wt. %$_{650 C}$: 8.4 | |
| $\Delta H_{vol}$-µv.sec/mg: 73 | |
| **Proximate analysis** | |
| Volatiles$_{Total}$-wt%: 95.5 | |
| Volatiles$_{650 C}$-wt%: 91.6 | |
| Volatiles$_{650-950 C}$-wt%: 3.9 | |
| Fixed carbon-wt%: 3.7 | |
| Ash-wt%: 0.8 | |

**Paper, Kimwipe tissue**

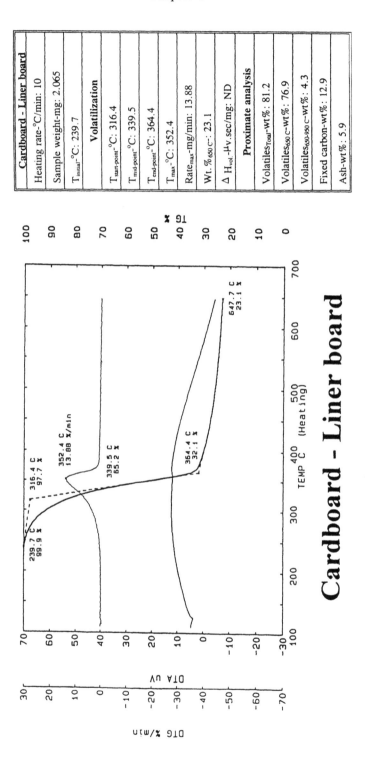

| Cardboard - Liner board | |
|---|---|
| Heating rate-°C/min: 10 | |
| Sample weight-mg: 2.065 | |
| $T_{initial}$-°C: 239.7 | |
| **Volatilization** | |
| $T_{start-point}$-°C: 316.4 | |
| $T_{mid-point}$-°C: 339.5 | |
| $T_{end-point}$-°C: 364.4 | |
| $T_{max}$-°C: 352.4 | |
| $Rate_{max}$-mg/min: 13.88 | |
| Wt.% $_{650\,C}$: 23.1 | |
| $\Delta H_{vol}$-μv.sec/mg: ND | |
| **Proximate analysis** | |
| $Volatiles_{Total}$-wt%: 81.2 | |
| $Volatiles_{650\,C}$-wt%: 76.9 | |
| $Volatiles_{650-950\,C}$-wt%: 4.3 | |
| Fixed carbon-wt%: 12.9 | |
| Ash-wt%: 5.9 | |

# Cardboard – Liner board

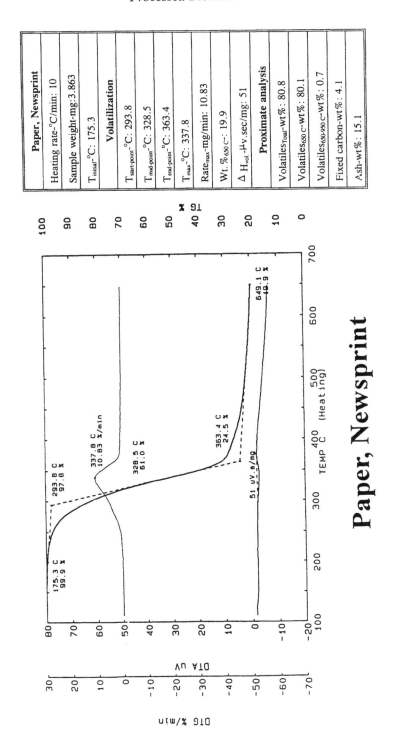

| Paper, Newsprint | |
|---|---|
| Heating rate-°C/min: 10 | |
| Sample weight-mg:3.863 | |
| $T_{initial}$-°C: 175.3 | |
| **Volatilization** | |
| $T_{start-point}$-°C: 293.8 | |
| $T_{mid-point}$-°C: 328.5 | |
| $T_{end-point}$-°C: 363.4 | |
| $T_{max}$-°C: 337.8 | |
| $Rate_{max}$-mg/min: 10.83 | |
| Wt.%$_{650}$C: 19.9 | |
| $\Delta H_{vol}$-$\mu$v.sec/mg: 51 | |
| **Proximate analysis** | |
| Volatiles$_{Total}$-wt%: 80.8 | |
| Volatiles$_{650}$c-wt%: 80.1 | |
| Volatiles$_{650-950}$c-wt%: 0.7 | |
| Fixed carbon-wt%: 4.1 | |
| Ash-wt%: 15.1 | |

**Paper, Newsprint**

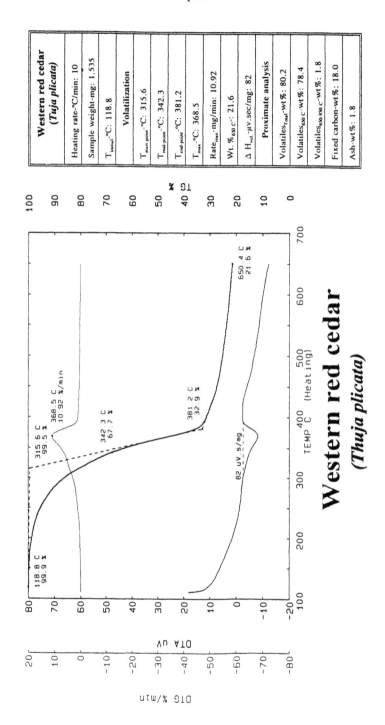

| Western red cedar (*Thuja plicata*) | |
|---|---|
| Heating rate-°C/min: 10 | |
| Sample weight-mg: 1.535 | |
| $T_{initial}$-°C: 118.8 | |
| **Volatilization** | |
| $T_{start\ point}$-°C: 315.6 | |
| $T_{mid\ point}$-°C: 342.3 | |
| $T_{end\ point}$-°C: 381.2 | |
| $T_{max}$-°C: 368.5 | |
| $Rate_{max}$-mg/min: 10.92 | |
| $Wt.\%_{650\ c}$: 21.6 | |
| $\Delta H_{vol}$-$\mu V.sec/mg$: 82 | |
| **Proximate analysis** | |
| $Volatiles_{Total}$-wt%: 80.2 | |
| $Volatiles_{650\ c}$-wt%: 78.4 | |
| $Volatiles_{650-950\ c}$-wt%: 1.8 | |
| Fixed carbon-wt%: 18.0 | |
| Ash-wt%: 1.8 | |

**Western red cedar**
*(Thuja plicata)*

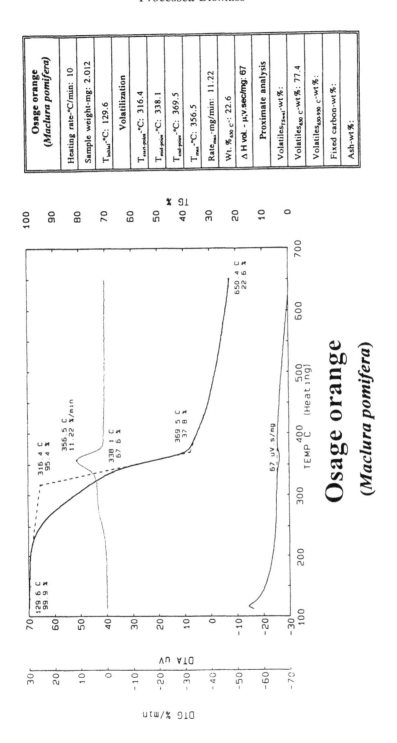

**Osage orange**
*(Maclura pomifera)*

| Osage orange *(Maclura pomifera)* |
| --- |
| Heating rate-°C/min: 10 |
| Sample weight-mg: 2.012 |
| $T_{initial}$-°C: 129.6 |
| **Volatilization** |
| $T_{start-point}$-°C: 316.4 |
| $T_{mid-point}$-°C: 338.1 |
| $T_{end-point}$-°C: 369.5 |
| $T_{max}$-°C: 356.5 |
| $Rate_{max}$-mg/min: 11.22 |
| Wt.%$_{650}$ c°: 22.6 |
| ΔH vol. - μ,v.sec/mg: 67 |
| **Proximate analysis** |
| Volatiles$_{TSwal}$-wt%: |
| Volatiles$_{650}$ c-wt%: 77.4 |
| Volatiles$_{650-950}$ c-wt%: |
| Fixed carbon-wt%: |
| Ash-wt%: |

# 9

# MUNICIPAL SOLID WASTES

## 9.1  MUNICIPAL SOLID WASTES AND THEIR DISTRIBUTION

Municipal solid waste (MSW) is composed of the discards from materials in daily use. There are three principal sources for this kind of waste:

1. Household
2. Construction and demolition
3. Commercial and industrial

Materials in municipal solid waste come from packaging devices such as canning and plastic bottles, used plastics from the medical and food industries, used paper, glass, and so on. Although the resource recovery from the entire municipal solid waste does not suggest a viable economical potential, certain components have received an encouraging industrial outlook. Technologies for aluminum and glass recycling are already developed to a level where they are of commercial benefit.

## 9.2  RECYCLING OF MUNICIPAL SOLID WASTES

The recycling of plastics is considered one of the most promising steps in the utilization of municipal solid wastes. Several technol-

**TABLE 9.1**

Types of Refuse-Derived Fuels

| RDF | Description |
|---|---|
| RDF-1 | Waste used as fuel in discarded form. |
| RDF-2 | Wastes processed to coarse particle size with or without ferrous metal separation. |
| RDF-3 | Shredded fuel derived from municipal solid waste that has been processed to remove metal, glass, and other inorganic materials. This material has a size smaller than 50 mm. |
| RDF-4 | Shredded combustible waste processed into powder form with size less than 2 mm. |
| RDF-5 | Combustible waste densified into pellets or briquettes. It is also known as d-RDF. |
| RDF-6 | Combustible waste processed into liquid fuel. |
| RDF-7 | Combustible waste processed into gaseous fuel. |

ogies are being developed to recycle plastics back into their monomers or to convert them into a viable energy source as an alternate fuel. In some cases, attempts are being made to produce some other manufacturing component from these plastic wastes, but their full exploitation at the commercial level still lies ahead.

Since most of the recycling technologies for plastics are based on thermal processes, one of the major steps is in the understanding of their thermal behavior. In this chapter, we present the thermograms of certain polymeric components of municipal solid waste, in an effort to have some insight in the thermal degradation of polymers in general. Some of these components form the part of *refuse-derived fuel* (RDF), which refers to a heterogeneous mixture of combustible materials, such as paper, plastic, and cardboards, that are separated from municipal solid waste. The term *refuse-derived fuel* was coined by Jerome Collins, who served with the New York State Environmental Facilities Corporation in 1973. The term *densified refuse-derived fuel* (d-RDF) was coined later by Harvey Alter, who served with the United States Chamber of Commerce. There are several kinds of refuse derived fuel, some of which are listed in Table 9.1.

## 9.3 METHODS FOR THE PREPARATION OF RDF

The major objective in the preparation of any fuel is in the maximizing of its calorific value and the minimization of ash content. In addition, during the preparation of RDF, extra care must be taken with regard to the removal of metallic and glass components that form the part of municipal solid waste. This process involves three major steps in the preparation of refuse-derived fuel, commonly referred to as the three *S*'s:

1. Sieving
2. Shredding
3. Screening

### 9.3.1 Sieving

*Sieving* helps in separating MSW into fractions of several sizes that are easier to handle, and helps set the stage for the separation of different components in the stream through the air classification process. Double sieving helps in removing fractions that are too small or too large in sizes. These fractions often cause major problems in fuel combustion in the furnace.

### 9.3.2 Shredding

*Shredding* helps in the handling of different types of MSW materials as part of the fuel preparation process that otherwise would have been discarded at the initial sieving stage. This stage helps in the maximization of the MSW usage on the quantitative basis. However, there are certain technologies at the shredding stage that can cause more harm than good. For example, shredding by the use of a fast milling process can imbed the inorganics in the fuel and increase the overall ash content. Hence, it is essential to select appropriate technologies during this stage of fuel preparation.

### 9.3.3 Screening

*Screening* refers to classification of fuel in fine gradations. This can be achieved either by air classification techniques such as cyclone separators or by laminar air classifiers, where very sharp cuts with reference to particle sizes are made with the help of trommels, which are rotating sieves, and help in the size gradation of the fuel. It is apparent that the classification by trommels is rather crude in nature when compared to air classification, but it is also cheaper. The judgment with regards to cost effectiveness and fuel quality is the one that makes the difference in these two approaches of screening techniques.

## 9.4  PREPARATION TECHNIQUES FOR DIFFERENT RDF TYPES

### 9.4.1  Preparation of RDF-1

There is no preparation technique for RDF-1; it is the use of MSW as a fuel source in as-is discarded form.

### 9.4.2  Preparation of RDF-2

During the preparation of this fuel, the MSW is first coarsely sieved and then shredded. In order to remove the metallic components of the MSW, a step for magnetic separation is included. In some cases, two stages of sieving have been utilized in order to better handle the fuel material. The RDF-2 fuel is a low-class fuel and is usually of high ash content.

### 9.4.3  Preparation of RDF-3

During the preparation of RDF-3 from MSW, all three steps (sieving, shredding, and screening) are utilized to maximize the fuel quality. At the sieving stage, large-size components are removed and the fuel is broadly classified into coarse and fine particles. In the second stage, the shredding of the large-size components take place, followed by the third stage, where shredded components are sent either through the sieving process once again or to the air classification system. However, it has been found that in most cases the exact step sequence has not provided the best results. Instead variations have been made to improve the process performance. For instance, there are one or two steps of sieving or shredding to improve the material handling and fuel quality in terms of ash removal and combustibility. It has been found that particles of larger sizes have a hard time during combustion, especially in a suspension boiler.

### 9.4.4  Preparation of RDF-4

Preparation of RDF-4 includes shredding followed by air classification and screening from trommel. This is followed by some chemical treatment, hot ball milling, and screening. This process was used in the preparation of *ecofuel*, for which the thermogram is provided later in this chapter. This cellulose-based fuel is an example in which major attempts were made to maximize the fuel quality derived from MSW. The process, however, has not proven economical up to this stage.

### 9.4.5    Preparation of RDF-5 or d-RDF

In the preparation of an RDF-5 or d-RDF type of fuel, the raw material is essentially the product of the steps involved in the production of RDF-3, which is then densified with the help of a suitable device such as extruder, impact briquette, or hot end extruder. However, special efforts should be made to remove the inorganic substances from the raw material in order to minimize the wear and tear of the densifier.

### 9.4.6    Preparation of RDF-6

The preparation of this type of fuel is done by selective pyrolysis of MSW under controlled conditions. The objective here is to maximize the conversion of solid combustible matter from the MSW into usable liquid fuel. One of the other techniques that is being currently pursued is the liquefaction of polymeric-based MSW to liquid fuels by the application of a liquefaction technique developed for coal. This technology exploits the concept of *hydrogen donor* to the fuel and converting the carbonaceous residue into hydrocarbon fuel.

### 9.4.7    Preparation of RDF-7

The preparation of this type of fuel takes the use of pyrolysis or gasification technology to convert the carbonaceous residue and the polymers into energy carrier gases, such as producer gas, or the hydrocarbon volatiles generated due to the thermal degradation under inert atmosphere. This process directly utilizes the thermal data presented in this book to understand the thermal behavior of plastics.

## 9.5    VARIATIONS IN RDF

Like any other fuel, the variability in the RDF is described on the basis of fuel properties such as calorific value, thermal degradation, and proximate and ultimate analysis. However, to have a better understanding, it is essential to compare these properties on moisture- and ash-free bases. In general, the combustible fraction of RDF is composed of cellulose-based materials such as paper and cardboards or polymer-based materials such as polyethylene (PE), polyethylene terephthalate (PET), or polytetrafluoroethylene (PTFE). If RDF primarily comprises cellulose-based material, it will have a low calorific value in the range of 13–20 MJ/kg; if the RDF is primarily polymer-based, then the calorific value will be on the higher end, i.e., around 35–60 MJ/kg. These values can then be adjusted based on the amount of ash and moisture content.

The quality of MSW from one geographic area was compared and studied in detail over a period of time by Kerklin et al. (1982). They found that variability in the heating value for MSW on the ash- and moisture-free basis was not that significant; rather it was very homogeneous, ranging from 18 to 23 MJ/kg. The heating values for five round robin samples of RDF-3 were in the range of 21–22 MJ/kg on a moisture- and ash-free basis. This shows that the quality of the fuel produced from MSW is very homogeneous in terms of its heating value. The effects of ash and moisture have demonstrated that these values can be significantly changed due to changes of ash and moisture content. The ash content for MSW available in the United States varied from 8 to 28%. This distribution in ash quantity is sufficient to change the thermal value of MSW-derived fuels.

The ultimate analysis of RDF-3 round robin samples is given in Table 9.2. It can be seen that the major combustible portion of the RDF is constituted by carbon, which is present in the range of 40–46%, with hydrogen being only in the range of 4–6% on a weight basis. This shows that the technology to liquefy RDF by using the hydrogen donor technique may well increase the overall fuel value of MSW; however, this technique is not proving to be economically feasible at this stage.

Another major aspect associated with the use of RDF as a fuel source is the apparent concern about the presence of chlorine, even though most of this chlorine is not present in the form that will promote corrosion or other health hazard. However, the limited amount that is present in the form that can promote these problems or lead to the liberation of chlorinated products of incomplete combustion (PICs) makes it imperative to address this aspect in much detail before the combustion of RDFs can be made feasible.

**TABLE 9.2**

Ultimate Analysis of RDF-3 Round Robin Samples

| Sample | Carbon wt% | Hydrogen wt% | Nitrogen wt% | Sulfur wt% | Chlorine wt% |
|--------|-----------|--------------|--------------|------------|--------------|
| 1 | 46.33 | 6.17 | 0.76 | 0.34 | 0.58 |
| 2 | 40.02 | 5.37 | 0.64 | 0.24 | 0.38 |
| 3 | 42.6 | 5.84 | 0.57 | 0.48 | 0.57 |

*Source*: Alter (1983).

## 9.6  THERMAL DEGRADATION AND MECHANISM

The thermal degradation pattern for RDF and its polymeric components can be understood with the help of thermograms.

### 9.6.1  Refuse-Derived Fuels

The thermograms of the three RDFs presented in this book show that the thermal degradation of RDF initiates at around 115–125°C and reaches to completion of 650°C. The presence of two distinct slopes indicate that the fuel is made up of two distinct groups of combustible materials. The first slope possibly indicates the cellulose-based materials, while the latter slope indicates the presence of polymer-based materials. The fixed carbon content lies in the range of 15–22% on a weight basis, while the total volatile is in the range of 70–80%. The ash content of these fuels in not that high (4–8%) when compared to the average value of 25–35% listed in the literature (Alter, 1983).

### 9.6.2  Polymer Components

The thermograms for polymers presented in this chapter show the individual thermal degradation behavior of some of the components often found in MSW. One clear distinction that can be made between polymers is that the vinyl-based polymers such as PVC degrade in two discrete steps, which is not always true for other polymeric materials. In addition, vinyl compounds tend to produce carbonaceous residue during thermal degradation while other polymers go through the degradation process by converting themselves completely into the volatile phase, with the exception of polyhydroxy benzoic acid. Therefore, in general, it can be said that most of the polymers undergo complete devolatilization with the generalized exception of vinyl compounds.

### 9.6.3  Degradation Process for Polymeric Components

*Polymethyl methacrylate* (PMMA) is considered the classical example of a polymer that degrades by reverse polymerization process. It has been found that the molecular weight of the polymer remains unaltered for the most part (75% polymer degradation) of the degradation process. The degradation of PMMA initiates at around 180°C.

Thermal degradation of *polystyrene* also initiates at 180°C, but significant conversion does not start before 350°C. Polystyrene has a number of oxygen groups distributed at random over the polymer chain, which acts as a weak spot and set of the degradation process. A typical product analysis from polystyrene degradation shows that about 90% by weight monomer (styrene) or dimer are distilled; the balance of 10% is distributed over ethane, toluene, benzene, and ethylbenzene.

The thermal degradation of polyethylene shows that very little monomer is generated. The product analysis shows that it mainly constitutes paraffins with up to 50% carbon atoms, with butenes, n-butane, propane, and ethane constituting about 75% of the volatile products. Polyethylene goes through a melt phase at around 150°C and then starts to degrade at 350°C.

*Polytetrafluoroethylene* degradation starts at around 400–450°C. The degradation process consists of stripping the monomers off the polymeric chain ends. The final product analysis shows the presence of $C_2F_4$ accounting for almost 95 wt % of the total polymer.

The thermal degradation process of *polyvinyl acetate* shows that the polymer is sometimes thermally stable up to 150°C and then starts to degrade liberating acetic acid, ketene, and carbon dioxide. The production of ketene and carbon dioxide is mainly due to the secondary degradation of acetic acid. It is suggested that the degradation process initiates by the liberation of one acetic acid molecule, which then leads to the formation of a double bond at the end of the polymer chain. This double bond travels the chain length and gives successive rise to the acetyl groups. Once all of the acetyl groups are liberated, the residue goes through the aromatization process. Upon further degradation, compounds such as benzene and toluene are liberated, resulting in about 90% or more polymer in the volatile state and the balance is left over as the carbonaceous residue.

*Polyvinyl chloride* degradation is similar to that of polyvinyl acetate in the sense that here the hydrochloric acid is at first liberated. The process initiates at around 200°C. The liberation of hydrochloric acid is, however, found to be due to structural weak spots in the chain. During the liberation of HCl, the polymer goes through an aromatization process that leads to the liberation of compounds such as benzene and toluene. The liberation of HCl accounts for almost 60–65 wt% of the PVC sample. The final carbonaceous residue is of the order of 3–7 wt %.

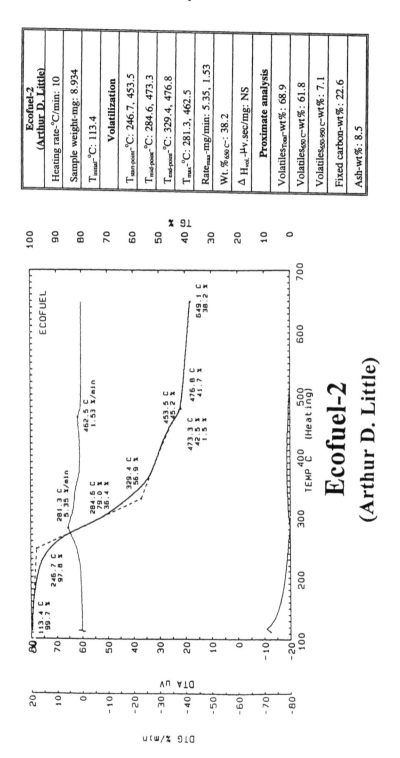

| Ecofuel-2 (Arthur D. Little) | |
|---|---|
| Heating rate-°C/min: 10 | |
| Sample weight-mg: 8.934 | |
| $T_{initial}$-°C: 113.4 | |
| **Volatilization** | |
| $T_{start-point}$-°C: 246.7, 453.5 | |
| $T_{mid-point}$-°C: 284.6, 473.3 | |
| $T_{end-point}$-°C: 329.4, 476.8 | |
| $T_{max}$-°C: 281.3, 462.5 | |
| $Rate_{max}$-mg/min: 5.35, 1.53 | |
| Wt. $\%_{650\,C}$: 38.2 | |
| $\Delta H_{vol.}$-$\mu$v.sec/mg: NS | |
| **Proximate analysis** | |
| Volatiles$_{Total}$-wt%: 68.9 | |
| Volatiles$_{650\,C}$-wt%: 61.8 | |
| Volatiles$_{650-950\,C}$-wt%: 7.1 | |
| Fixed carbon-wt%: 22.6 | |
| Ash-wt%: 8.5 | |

# Ecofuel-2
## (Arthur D. Little)

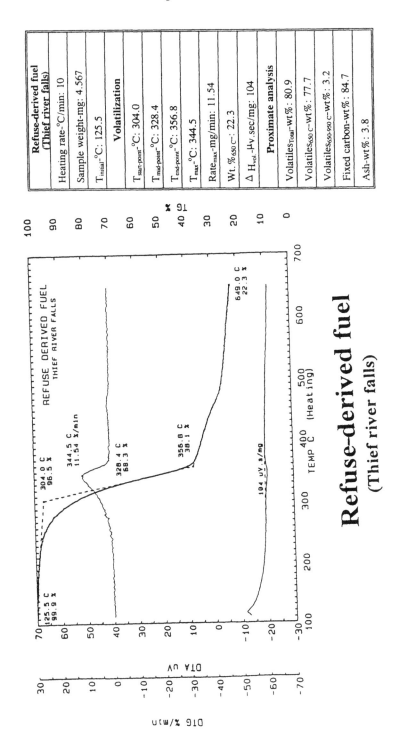

| Refuse-derived fuel (Thief river falls) | |
| --- | --- |
| Heating rate-°C/min: 10 | |
| Sample weight-mg: 4.567 | |
| $T_{initial}$-°C: 125.5 | |
| **Volatilization** | |
| $T_{start-point}$-°C: 304.0 | |
| $T_{mid-point}$-°C: 328.4 | |
| $T_{end-point}$-°C: 356.8 | |
| $T_{max}$-°C: 344.5 | |
| $Rate_{max}$-mg/min: 11.54 | |
| Wt.%$_{650\,C}$: 22.3 | |
| $\Delta H_{vol}$-$\mu$v.sec/mg: 104 | |
| **Proximate analysis** | |
| $Volatiles_{Total}$-wt%: 80.9 | |
| $Volatiles_{650\,C}$-wt%: 77.7 | |
| $Volatiles_{650-950\,C}$-wt%: 3.2 | |
| Fixed carbon-wt%: 84.7 | |
| Ash-wt%: 3.8 | |

**Refuse-derived fuel**
(Thief river falls)

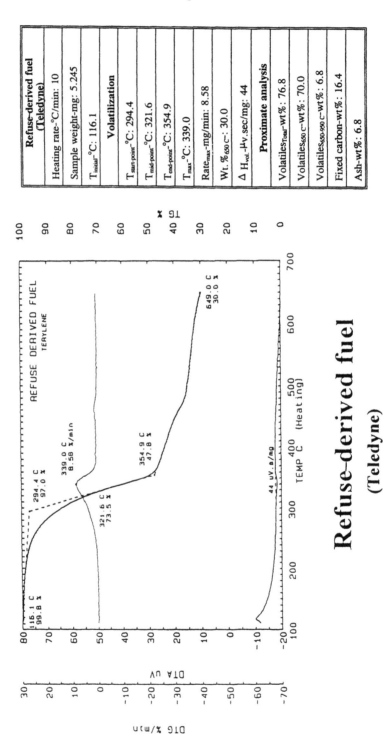

| Refuse-derived fuel (Teledyne) | |
|---|---|
| Heating rate-°C/min: | 10 |
| Sample weight-mg: | 5.245 |
| $T_{initial}$-°C: | 116.1 |
| **Volatilization** | |
| $T_{start-point}$-°C: | 294.4 |
| $T_{mid-point}$-°C: | 321.6 |
| $T_{end-point}$-°C: | 354.9 |
| $T_{max}$-°C: | 339.0 |
| $Rate_{max}$-mg/min: | 8.58 |
| Wt.%$_{650 C}$: | 30.0 |
| $\Delta H_{vol}$-$\mu$v.sec/mg: | 44 |
| **Proximate analysis** | |
| Volatiles$_{Total}$-wt%: | 76.8 |
| Volatiles$_{650 C}$-wt%: | 70.0 |
| Volatiles$_{650-990 C}$-wt%: | 6.8 |
| Fixed carbon-wt%: | 16.4 |
| Ash-wt%: | 6.8 |

# Refuse-derived fuel
## (Teledyne)

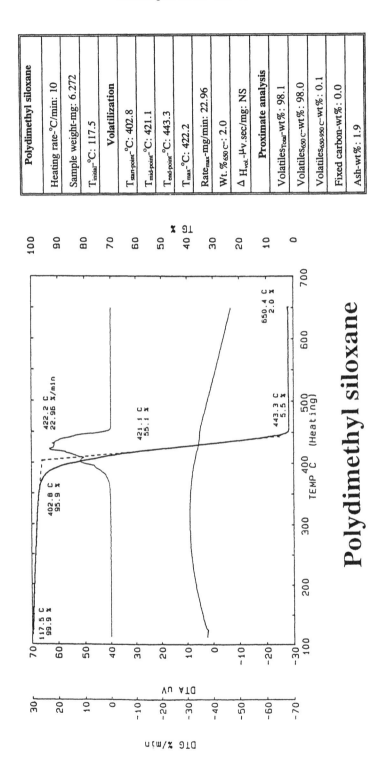

## Polydimethyl siloxane

| Polydimethyl siloxane | |
|---|---|
| Heating rate-°C/min: 10 | |
| Sample weight-mg: 6.272 | |
| $T_{initial}$-°C: 117.5 | |
| **Volatilization** | |
| $T_{start-point}$-°C: 402.8 | |
| $T_{mid-point}$-°C: 421.1 | |
| $T_{end-point}$-°C: 443.3 | |
| $T_{max}$-°C: 422.2 | |
| $Rate_{max}$-mg/min: 22.96 | |
| Wt.%$_{650\,C}$: 2.0 | |
| $\Delta H_{vol.}$-$\mu$v.sec/mg: NS | |
| **Proximate analysis** | |
| Volatiles$_{Total}$-wt%: 98.1 | |
| Volatiles$_{650\,C}$-wt%: 98.0 | |
| Volatiles$_{650-950\,C}$-wt%: 0.1 | |
| Fixed carbon-wt%: 0.0 | |
| Ash-wt%: 1.9 | |

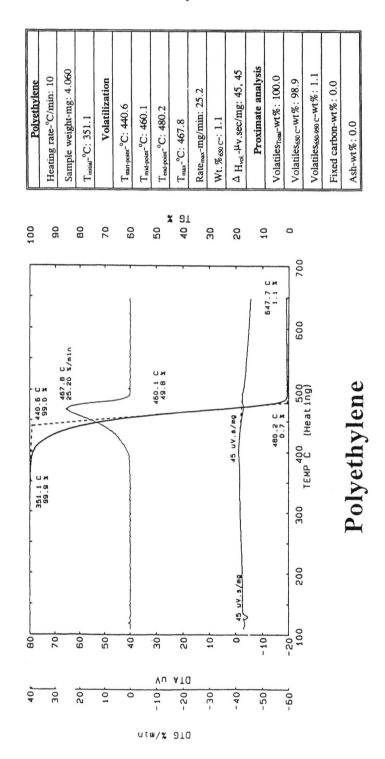

| Polyethylene | |
|---|---|
| Heating rate-°C/min: 10 | |
| Sample weight-mg: 4.060 | |
| $T_{initial}$-°C: 351.1 | |
| **Volatilization** | |
| $T_{start-point}$-°C: 440.6 | |
| $T_{mid-point}$-°C: 460.1 | |
| $T_{end-point}$-°C: 480.2 | |
| $T_{max}$-°C: 467.8 | |
| $Rate_{max}$-mg/min: 25.2 | |
| Wt.%$_{650\,C}$: 1.1 | |
| $\Delta H_{vol.}$-$\mu v.sec/mg$: 45, 45 | |
| **Proximate analysis** | |
| Volatiles$_{Total}$-wt%: 100.0 | |
| Volatiles$_{650\,C}$-wt%: 98.9 | |
| Volatiles$_{650-950\,C}$-wt%: 1.1 | |
| Fixed carbon-wt%: 0.0 | |
| Ash-wt%: 0.0 | |

# Polyethylene

192

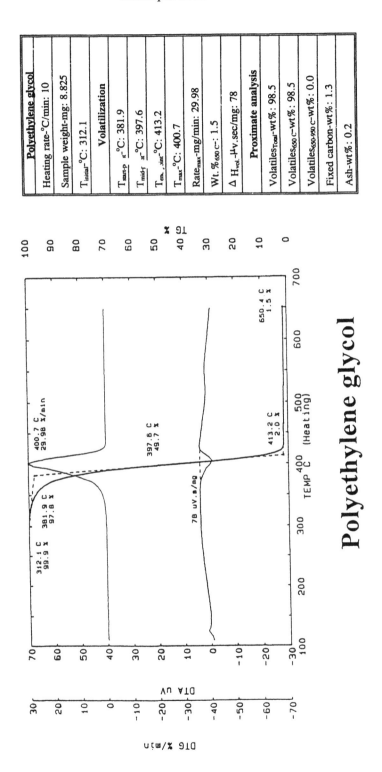

| Polyethylene glycol | |
|---|---|
| Heating rate-°C/min: 10 | |
| Sample weight-mg: 8.825 | |
| $T_{initial}$-°C: 312.1 | |
| **Volatilization** | |
| $T_{start-p}$-°C: 381.9 | |
| $T_{mid-r}$-°C: 397.6 | |
| $T_{en,\_int}$-°C: 413.2 | |
| $T_{max}$-°C: 400.7 | |
| $Rate_{max}$-mg/min: 29.98 | |
| Wt.%$_{650C}$: 1.5 | |
| $\Delta H_{vol}$-Hv.sec/mg: 78 | |
| **Proximate analysis** | |
| Volatiles$_{Total}$-wt%: 98.5 | |
| Volatiles$_{650C}$-wt%: 98.5 | |
| Volatiles$_{650-950C}$-wt%: 0.0 | |
| Fixed carbon-wt%: 1.3 | |
| Ash-wt%: 0.2 | |

# Polyethylene glycol

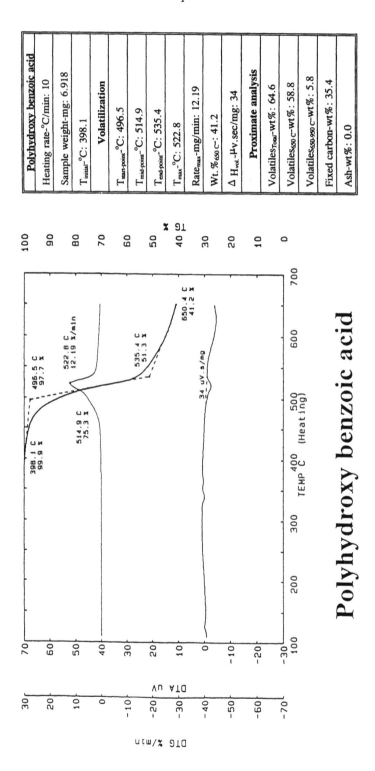

| Polyhydroxy benzoic acid | |
|---|---|
| Heating rate-°C/min: | 10 |
| Sample weight-mg: | 6.918 |
| $T_{initial}$-°C: | 398.1 |
| **Volatilization** | |
| $T_{start-point}$-°C: | 496.5 |
| $T_{mid-point}$-°C: | 514.9 |
| $T_{end-point}$-°C: | 535.4 |
| $T_{max}$-°C: | 522.8 |
| Rate$_{max}$-mg/min: | 12.19 |
| Wt.%$_{650°C}$: | 41.2 |
| $\Delta H_{vol}$-$\mu$v.sec/mg: | 34 |
| **Proximate analysis** | |
| Volatiles$_{Total}$-wt%: | 64.6 |
| Volatiles$_{650°C}$-wt%: | 58.8 |
| Volatiles$_{650-950°C}$-wt%: | 5.8 |
| Fixed carbon-wt%: | 35.4 |
| Ash-wt%: | 0.0 |

# Polyhydroxy benzoic acid

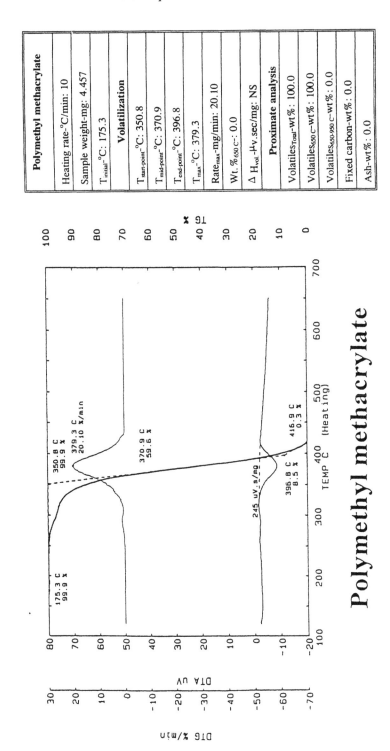

| Polymethyl methacrylate | |
|---|---|
| Heating rate-°C/min: | 10 |
| Sample weight-mg: | 4.457 |
| $T_{initial}$-°C: | 175.3 |
| **Volatilization** | |
| $T_{start-point}$-°C: | 350.8 |
| $T_{mid-point}$-°C: | 370.9 |
| $T_{end-point}$-°C: | 396.8 |
| $T_{max}$-°C: | 379.3 |
| $Rate_{max}$-mg/min: | 20.10 |
| Wt.%$_{650 C}$: | 0.0 |
| $\Delta H_{vol}$-$\mu$v.sec/mg: | NS |
| **Proximate analysis** | |
| Volatiles$_{Total}$-wt%: | 100.0 |
| Volatiles$_{650 C}$-wt%: | 100.0 |
| Volatiles$_{650-950 C}$-wt%: | 0.0 |
| Fixed carbon-wt%: | 0.0 |
| Ash-wt%: | 0.0 |

## Polymethyl methacrylate

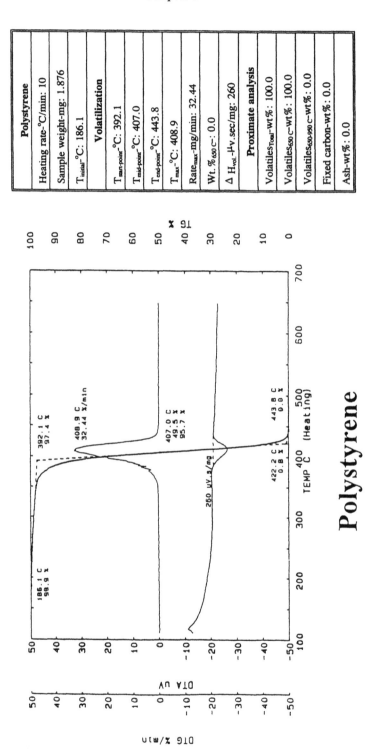

| Polystyrene | |
| --- | --- |
| Heating rate-°C/min: | 10 |
| Sample weight-mg: | 1.876 |
| $T_{initial}$-°C: | 186.1 |
| **Volatilization** | |
| $T_{start-point}$-°C: | 392.1 |
| $T_{mid-point}$-°C: | 407.0 |
| $T_{end-point}$-°C: | 443.8 |
| $T_{max}$-°C: | 408.9 |
| Rate$_{max}$-mg/min: | 32.44 |
| Wt. %$_{650C}$: | 0.0 |
| $\Delta H_{vol.}$-$\mu$v.sec/mg: | 260 |
| **Proximate analysis** | |
| Volatiles$_{Total}$-wt%: | 100.0 |
| Volatiles$_{650C}$-wt%: | 100.0 |
| Volatiles$_{650-990C}$-wt%: | 0.0 |
| Fixed carbon-wt%: | 0.0 |
| Ash-wt%: | 0.0 |

## Polystyrene

| Polytetrafluoroethylene (Teflon) | |
| --- | --- |
| Heating rate-°C/min: 10 | |
| Sample weight-mg: 3.076 | |
| $T_{initial}$-°C: 447.8 | |
| **Volatilization** | |
| $T_{start-point}$-°C: 568.0 | |
| $T_{mid-point}$-°C: 586.9 | |
| $T_{end-point}$-°C: 605.8 | |
| $T_{max}$-°C: 592.7 | |
| $Rate_{max}$-mg/min: 25.8 | |
| Wt.%$_{650°C}$: 0.5 | |
| $\Delta H_{vol}$-$\mu$v.sec/mg: 11 | |
| **Proximate analysis** | |
| Volatiles$_{TSwa}$-wt%: 100.0 | |
| Volatiles$_{650C}$-wt%: 99.5 | |
| Volatiles$_{650-990C}$-wt%: 0.5 | |
| Fixed carbon-wt%: 0.0 | |
| Ash-wt%: 0.0 | |

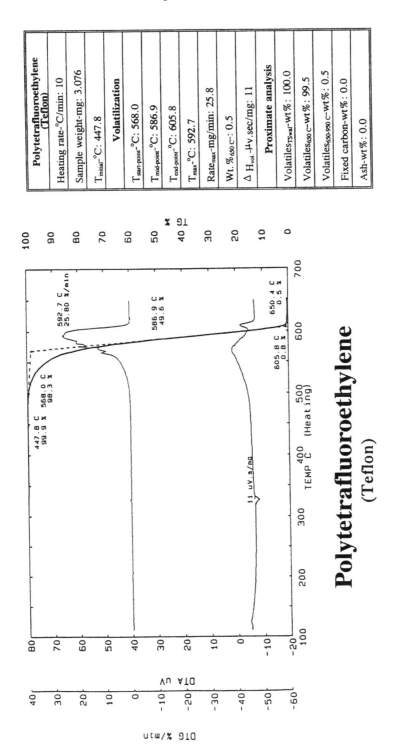

## Polytetrafluoroethylene
### (Teflon)

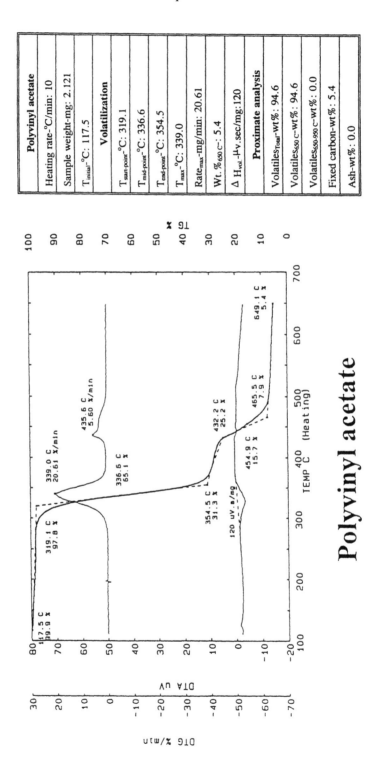

| Polyvinyl acetate | |
|---|---|
| Heating rate-°C/min: 10 | |
| Sample weight-mg: 2.121 | |
| $T_{initial}$-°C: 117.5 | |
| **Volatilization** | |
| $T_{start-point}$-°C: 319.1 | |
| $T_{mid-point}$-°C: 336.6 | |
| $T_{end-point}$-°C: 354.5 | |
| $T_{max}$-°C: 339.0 | |
| Rate$_{max}$-mg/min: 20.61 | |
| Wt. %$_{650 C}$-: 5.4 | |
| $\Delta H_{vol}$-J-v.sec/mg:120 | |
| **Proximate analysis** | |
| Volatiles$_{Total}$-wt%: 94.6 | |
| Volatiles$_{650 C}$-wt%: 94.6 | |
| Volatiles$_{650-950 C}$-wt%: 0.0 | |
| Fixed carbon-wt%: 5.4 | |
| Ash-wt%: 0.0 | |

## Polyvinyl acetate

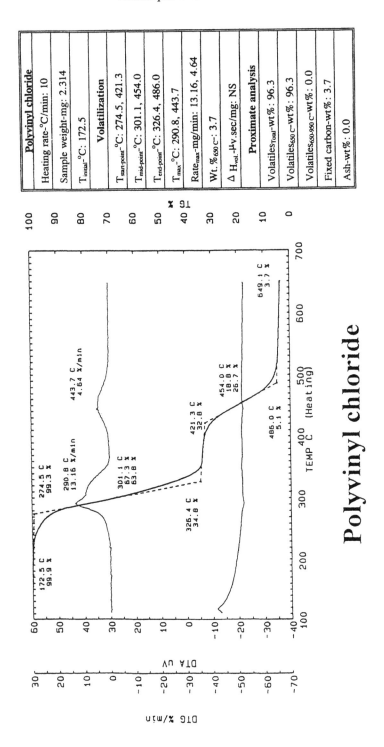

| Polyvinyl chloride | |
| --- | --- |
| Heating rate-°C/min: | 10 |
| Sample weight-mg: | 2.314 |
| $T_{initial}$-°C: | 172.5 |
| **Volatilization** | |
| $T_{start-point}$-°C: | 274.5, 421.3 |
| $T_{mid-point}$-°C: | 301.1, 454.0 |
| $T_{end-point}$-°C: | 326.4, 486.0 |
| $T_{max}$-°C: | 290.8, 443.7 |
| $Rate_{max}$-mg/min: | 13.16, 4.64 |
| Wt. %$_{650°C}$: | 3.7 |
| $\Delta H_{vol.}$-µv.sec/mg: | NS |
| **Proximate analysis** | |
| Volatiles$_{Total}$-wt%: | 96.3 |
| Volatiles$_{650°C}$-wt%: | 96.3 |
| Volatiles$_{650-950°C}$-wt%: | 0.0 |
| Fixed carbon-wt%: | 3.7 |
| Ash-wt%: | 0.0 |

## Polyvinyl chloride

199

# 10

## High-Carbon Solid Fuels

### 10.1 INTRODUCTION

There are two types of high-carbon fuels: some are naturally occurring, like coal; others are transformed by human activity, like charcoal, for which woody biomass is subjected to thermal treatment for carbon enrichment.

The origin of naturally occurring high-carbon solid fuels can be traced down to the biomass materials of previous years. It is said that nature feeds on itself, and that if it is left undisturbed the biomass plantations would eventually destroy themselves to the basic elements. However, sometimes due to certain interventions such as land form changes, earthquakes, and water sedimentation, these biomass plantations are buried and are then prevented from complete destruction, or better stated, their degradation process is intervened at an intermediate step. These plantations over the years alter under pressure, heat, and chemical interactions to produce a wide variety of naturally occurring high-carbon solid fuels like peat, coal, and natural waxes. The term *coal* is very generic in nature and is applied to a wide variety of carbonaceous fuels including lignite and peat.

Other than the genetic nature of the plant origin, the characteristics of coal are defined by two major aspects:

1.  The type and extent of biochemical activity
2.  Thermal alteration of the original mass

The coalification or the progressive enrichment of organically bound carbon has been used to classify coal in the following major groups:

Peat → lignite → subbituminous → bituminous anthracite

This classification, however, in no way relates to any genetic connection between the coal seams found under the same class. It merely relates to the extent of carbon presence or the extent to which the transformation of the original biomass has occurred. At this stage, it is also necessary to point out that the mere passage of time does not guarantee that bitiminous coal or any other intermediate form will transform to anthracite. It may well be that the degradation process has taken a path that does not permit further carbon enrichment.

## 10.2 PHYSICAL AND CHEMICAL COMPOSITION OF COAL

The physical designation of coal is based on banded components known as lithotypes:

*Vitrain* is a narrow, black, glossy band fractured into small cubes that appear structureless on the first glance.
*Clarain* is made up of horizontally striated black layers with silky luster.
*Durain* also occurs in layers, but the texture is tight and granular. In addition, the color of this coal is more gray than black.
*Fusain* is soft and fragile and breaks readily into powder form. Its texture can be compared more with charcoal than any other naturally occurring coal seams.

The International Committee for Coal Petrography (ICCP) has codified these various coal series into three maceral groups: vitrinite, exinite, and inertinite, which are further divided into several mocrolithotype groups. Vitrinites are generally composed of wood, bark, and humic gels. The extinites are composed of fungal spores, leafs waxes, and algal remains; and the inertinites are derived from grains and unspecified detrital matter.

As for most fuels, the chemical composition of coal is defined on the basis of proximate analysis, which provides a rough measure of the distribution of products in terms of moisture content, volatiles, fixed carbon content, and inorganic ash. The other measure, known as ultimate analysis, provides the elemental distribution of carbon, hydrogen, and oxygen in a solid fuel.

The moisture content of coal and other high-carbon solid fuels is determined by ASTM standard D 3173-73;30. The volatile matter is determined by ASTM D 3175-77. The inorganic ash is determined by the ASTM D 3174 test method. The fixed carbon for a sample is determined by difference, as shown below:

%FC = 100 − (% water + % volatile matter + % ash)

The ultimate analysis is performed by involving oxidation, decomposition, and reduction methods. The quantification of carbon and hydrogen is done by subjecting the sample to complete combustion to products $CO_2$ and water at 900°C. The exit stream is then purified of any oxides and analyzed to determine the total carbon and hydrogen content. The determination of nitrogen is done by digesting the coal sample with sulfuric acid and potassium sulfate in the presence of a catalyst like mercury. The cooled solution is then made alkaline and liberated ammonia is distilled to determine the total nitrogen content. The determination of sulfur is based on the combustion of sulfur-bearing compounds to sulfate ions, which are then measured volumetrically. The oxygen in these samples is determined by difference, as shown below:

% oxygen = 100 − (%C + %H + %N + %S)

The representative samples of coal studied in this book were obtained from the Argonne National Laboratory, Argonne, IL, and illustrate the variability in their thermal degradation pattern, volatile content, fixed carbon, and ash.

## 10.3   SOME GENERAL FACTS ABOUT COAL CONSTITUTION AND ITS CHEMISTRY

1. Coal is not completely aromatic in nature.
2. Its aromaticity varies with rank and ranges from 40% to 100%.
3. The high aromaticity in coal products is mainly due to secondary effects of conversion processes and not because of the intrinsic property.
4. Coal structure contains predominantly mono- and diaromatic rings followed by the presence of polycyclic aromatic rings.
5. The subbituminous coals have a significant presence of polycyclic aliphatic rings.

6. Coal is extremely reactive in the presence of hydrogen donors.

## 10.4  PYROLYSIS OF HIGH-CARBON FUELS

Most high-carbon fuels have a pyrolysis mechanism similar to that of coal. Since a wide variety of research work is available on coal pyrolysis, we believe it would be in the best interest of readers to explain the pyrolysis process for coal, which can then be extrapolated for majority of high-carbon fuels.

When coal samples are subjected to thermal degradation under inert atmosphere, they usually do not liberate any volatiles before 300°C except for the liberation of moisture content. This is in contrast to wood and other biomass samples, where the devolatilization process in some cases takes place at temperatures as low as 150°C. This difference is primarily due to the fact the coal has already gone through an in situ devolatilization process through compaction and thermal treatment.

The chemically combined water is liberated at around 150–250°C. At temperatures higher than 300°C, the structure of coal experiences a fragmentation process. During this stage, it passes through the plastic state (350–450°C), thereby forming a porous structure. It has been noted that the majority of volatile products (75%) are liberated under this temperature region; however, the product composition and the solid residue can be markedly different based on the thermal cycle, heating rate, and pressure conditions. This liberation of volatiles ends in the temperature range of 550–650°C. Further heating of the coal sample leads to secondary degasification, where the process of gradual elimination of heteroatoms, principally H and O, takes place. This process takes up to 850°C and in some cases is also carried out until 1000°C. The principal products of this stage of thermal degradation are water and oxides of carbon and hydrogen, among others. The thermal treatment of coal at this temperature range is also responsible for the aromatization process in the coal structure.

Variables that affect the thermal degradation of coal include temperature, pressure, particle size, residence time, and coal type. Among these, the singularly most important parameter is the final temperature at which the degradation process has been carried out, followed by residence time. Some of the common observations made during coal pyrolysis are:

1. The extent to which coal devolatilizes is highly dependent on the final temperature of pyrolysis.

2. The residence time required for the devolatilization of similar coal samples varies greatly with the change in particle size.

3. Coal samples of different chemical characteristics can have similar devolatilization characteristics.

4. Bituminous coals tend to liberate more liquid products while lignite tends to liberate more gaseous products.

5. The properties of final devolatilization products change with the change in final temperature for pyrolysis and the heat rate for pyrolysis.

Since the nature and amount of volatile matter depends on coal type, there is a direct relationship between the carbon content of the coal and the extent to which it decomposes at any particular temperature. It has been found that as the carbon content of the fuel increases, the devolatilization process takes place at relatively higher temperatures and the maximum weight loss rates become smaller.


## 10.4.1  Kinetics of Pyrolysis

Kinetic rate data for coal pyrolysis have usually been found by the utilization of constant heating rate thermogravimetric data. It is important to emphasize once again that the kinetic rate data on this technique are based on the assumption that the rate of decomposition is instantaneously reflected by the weight loss of the sample. Based on this assumption, the kinetic functions described in Chapter 3 can be applied to determine the rate parameters. Using this approach, van Krevelen et al. (1951) found that the thermal decomposition of coal follows a first-order reaction with activation energy in the range of 105–130 kJ/mole and the decomposition rates are controlled by the rupture of various covalent bonds. The qualitative description of the pyrolysis mechanism they proposed has been universally accepted, but the quantitative measurement of activation energy has been challenged on the grounds that there is a large built-in factor of diffusion resistance for the volatiles in several coal samples, and therefore the activation energy reported by this work is not the true indication of the pyrolysis process.

Before we dwell more on this aspect, it is imperative that we make a distinction between devolatilization, pyrolysis, and carbonization processes, as these words are interchangeably used in the literature without taking adequate care. The devolatilization process is the thermal degradation of a sample in which the intention is to liberate the volatile components of a solid sample. This process involves the pyrolysis process, which is defined as the extensive thermal degradation of the solid sample by which the volatile components are severed from the fixed carbon content of the sample, and to achieve the devolatilization process these volatiles need to be boiled off from the solid residue. This process would involve the kinetics of diffusion process and the distillation of the various volatile fractions. The carbonization process is the progressive enrichment of the solid carbon and usually takes place at temperatures higher than the pyrolysis process (>600°C). During this process, the oxides of carbon and hydrogen are liberated in an attempt to maximize the carbon content in the solid residue. However, if one desires, the process of carbon enrichment can be stopped by an intermediate step based on the desired final combustion properties of solid carbon.

Now, if we review the results obtained by van Krevelen et al. (1951) in the light of the above definitions, then it is found that the rate of pyrolysis has an apparent activation energy of 12–15 kJ/mole while the activation energy for the overall devolatilization process is in the range of 100–130 kJ/mole. These results are in agreement with the overall devolatilization measurements made by Kobayashi et al. (1976) on certain lignite and bituminous coals. The activation energy for lignite samples has been reported in the range of 100–110 kJ/mole, while for bituminous coals the activation energy is around 140–170 kJ/mole. The pre-exponential factor for lignite is $2 \times 10^5$ sec$^{-1}$ and that for bituminous coal it is $1.3 \times 10^7$ sec$^{-1}$.

## 10.5  CHARCOAL

If wood and other biomass materials are considered as humankind's earliest fuel, then the oldest synthetic fuel is charcoal. The use of wood for fuel produces a smoky, low-efficiency fire unless it is burned in well-designed stoves and furnaces. On the other hand, charcoal produces a very clean and hot fire needed for indoor cooking and metal working, and now used in many chemical processes. Conversion of wood to charcoal is a very old art. In early times, charcoal was made by the slow heating of wood in the absence of air (Earl, 1974). More recently, processes have been developed to make pyrolysis oils and charcoal simultaneously (Diebold and Schahill, 1988; Richard and Antal, 1994).

The name *charcoal* does not adequately describe the hundreds of varieties of materials in commercial use. One simple index of charcoals is the percent of volatiles remaining after pyrolysis. A very high volatile charcoal called torrefied wood is produced in the range 230–270°C. It is claimed that torrefied wood contains most of the fuel value of the original wood in a much higher concentration.

If wood is pyrolyzed in the range 300–360°C, then hemicellulose is selectively destroyed while the strength and cellular structure of the wood is preserved. This material is rendered hydrophobic and oleophilic and is an excellent absorbent for oils and chemicals marketed under the name *Sea Sweep* (Reed et al., 1994).

If wood is heated to about 300°C, the reaction becomes exothermic and can go spontaneously to about 450°C. This composition would represent the "cooking charcoal" still widely used around the world and responsible for much deforestation. While most easily made from wood, cooking charcoal can also be made from rice hulls, bagasse, and other waste materials if it is briquetted. Such processing could reduce the deforestation currently a subject of much concern.

Charcoal for metallurgical purposes typically is produced by heating above 450°C. It has a very low sulfur content and is used in making high-grade steel, among other things.

Finally, if charcoal is heated to 800–900°C, or chemically treated, an *activated charcoal* results that has a very high internal surface area. Typically only 20–40% of the original biomass remains, so that the density is very low. Activated charcoal is useful for absorbing large molecules, toxics, and so forth. Activated charcoals for liquid purification (i.e., sugar decolorizing) are used in small particle size. For gas purification, large particles are required and so the original biomass must be very dense in order to maintain the physical structure. Samples like coconut hulls are good for this purpose.

The thermograms of charcoal presented in this book show the extent of devolatilization and thus indicate the possible degree of charring.

## 10.6  PEAT

Like charcoal, peat has been burned as a fuel and used for fertilizer for thousands of years. It is burned commercially to produce power, particularly in Ireland and Finland. Peat is the remains of plants that have decomposed, usually under water. The peat must be dried before use because of its high moisture content. The International Humic Institute (IHI), located in Colorado, collects and analyzes various forms of peat and has contributed the samples studied here.

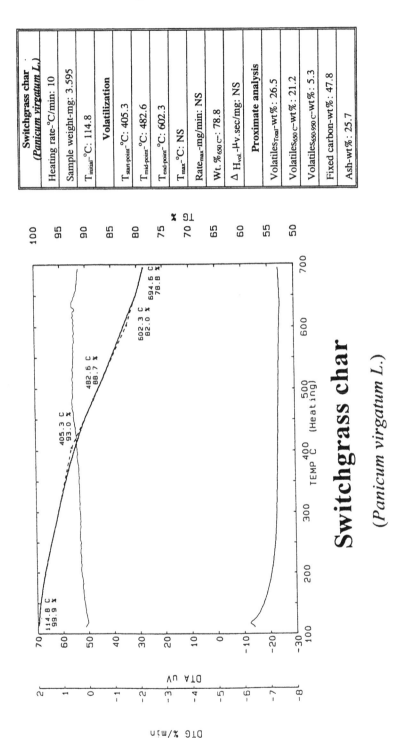

| Switchgrass char *(Panicum virgatum L.)* | |
|---|---|
| Heating rate–°C/min: 10 | |
| Sample weight–mg: 3.595 | |
| $T_{initial}$–°C: 114.8 | |
| **Volatilization** | |
| $T_{start-point}$–°C: 405.3 | |
| $T_{mid-point}$–°C: 482.6 | |
| $T_{end-point}$–°C: 602.3 | |
| $T_{max}$–°C: NS | |
| $Rate_{max}$–mg/min: NS | |
| Wt.%$_{650 C}$: 78.8 | |
| $\Delta H_{vol}$–$\mu$v.sec/mg: NS | |
| **Proximate analysis** | |
| Volatiles$_{Total}$–wt%: 26.5 | |
| Volatiles$_{650 C}$–wt%: 21.2 | |
| Volatiles$_{650-950 C}$–wt%: 5.3 | |
| Fixed carbon–wt%: 47.8 | |
| Ash–wt%: 25.7 | |

## Switchgrass char

*(Panicum virgatum L.)*

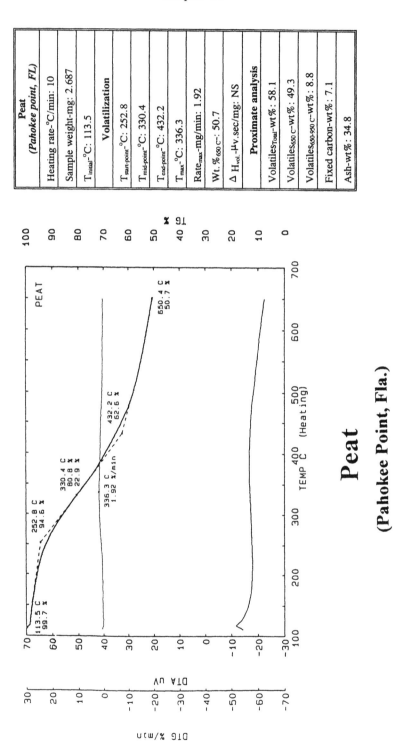

| Peat (Pahokee point, FL) | |
|---|---|
| Heating rate–°C/min: 10 | |
| Sample weight–mg: 2.687 | |
| $T_{initial}$–°C: 113.5 | |
| **Volatilization** | |
| $T_{start-point}$–°C: 252.8 | |
| $T_{mid-point}$–°C: 330.4 | |
| $T_{end-point}$–°C: 432.2 | |
| $T_{max}$–°C: 336.3 | |
| $Rate_{max}$–mg/min: 1.92 | |
| Wt.%$_{650 C}$: 50.7 | |
| $\Delta H_{vol}$–$\mu$v.sec/mg: NS | |
| **Proximate analysis** | |
| Volatiles$_{Total}$–wt%: 58.1 | |
| Volatiles$_{650 C}$–wt%: 49.3 | |
| Volatiles$_{650-950 C}$–wt%: 8.8 | |
| Fixed carbon–wt%: 7.1 | |
| Ash–wt%: 34.8 | |

# Peat
## (Pahokee Point, Fla.)

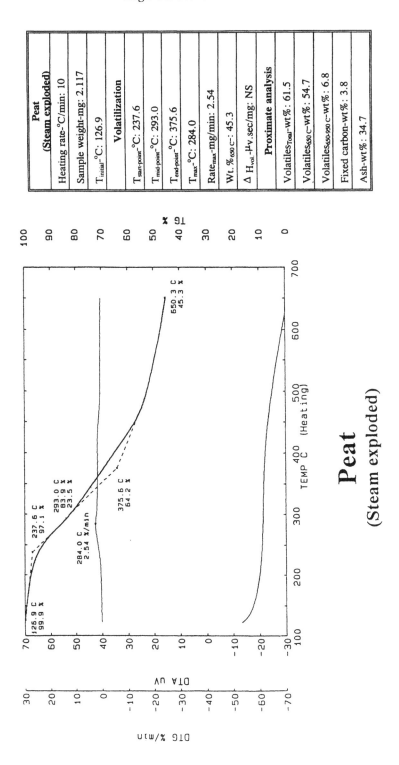

| Peat (Steam exploded) | |
|---|---|
| Heating rate-°C/min: | 10 |
| Sample weight-mg: | 2.117 |
| $T_{initial}$-°C: | 126.9 |
| **Volatilization** | |
| $T_{start-point}$-°C: | 237.6 |
| $T_{mid-point}$-°C: | 293.0 |
| $T_{end-point}$-°C: | 375.6 |
| $T_{max}$-°C: | 284.0 |
| $Rate_{max}$-mg/min: | 2.54 |
| Wt.%$_{650C}$: | 45.3 |
| $\Delta H_{vol}$.-μv.sec/mg: | NS |
| **Proximate analysis** | |
| Volatiles$_{Total}$-wt%: | 61.5 |
| Volatiles$_{650C}$-wt%: | 54.7 |
| Volatiles$_{650-990C}$-wt%: | 6.8 |
| Fixed carbon-wt%: | 3.8 |
| Ash-wt%: | 34.7 |

**Peat**
**(Steam exploded)**

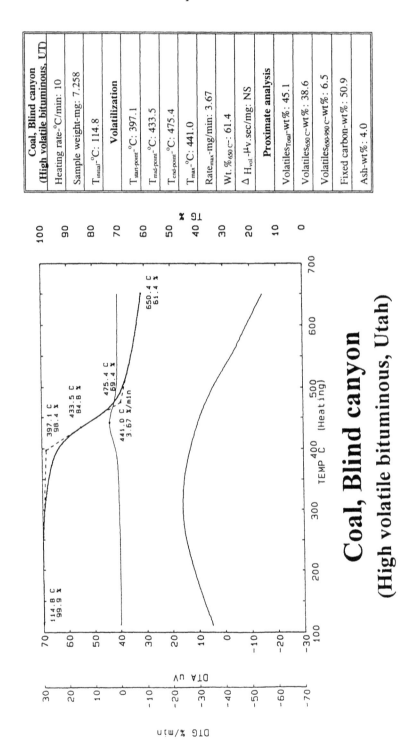

| Coal, Blind canyon (High volatile bituminous, UT) | |
|---|---|
| Heating rate-°C/min: | 10 |
| Sample weight-mg: | 7.258 |
| $T_{initial}$-°C: | 114.8 |
| **Volatilization** | |
| $T_{start-point}$-°C: | 397.1 |
| $T_{mid-point}$-°C: | 433.5 |
| $T_{end-point}$-°C: | 475.4 |
| $T_{max}$-°C: | 441.0 |
| $Rate_{max}$-mg/min: | 3.67 |
| Wt.%$_{650 C}$-: | 61.4 |
| $\Delta H_{vol}$-$\mu$v.sec/mg: | NS |
| **Proximate analysis** | |
| Volatiles$_{Total}$-wt%: | 45.1 |
| Volatiles$_{650 C}$-wt%: | 38.6 |
| Volatiles$_{650-950 C}$-wt%: | 6.5 |
| Fixed carbon-wt%: | 50.9 |
| Ash-wt%: | 4.0 |

# Coal, Blind canyon
## (High volatile bituminous, Utah)

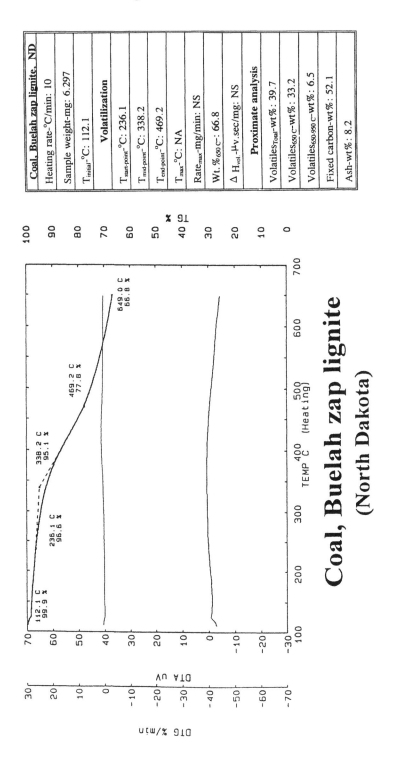

| Coal, Buelah zap lignite, ND | |
|---|---|
| Heating rate-°C/min: 10 | |
| Sample weight-mg: 6.297 | |
| $T_{initial}$-°C: 112.1 | |
| **Volatilization** | |
| $T_{start-point}$-°C: 236.1 | |
| $T_{mid-point}$-°C: 338.2 | |
| $T_{end-point}$-°C: 469.2 | |
| $T_{max}$-°C: NA | |
| $Rate_{max}$-mg/min: NS | |
| Wt.%$_{650C}$: 66.8 | |
| $\Delta H_{vol}$-µv.sec/mg: NS | |
| **Proximate analysis** | |
| Volatiles$_{Toal}$-wt%: 39.7 | |
| Volatiles$_{650C}$-wt%: 33.2 | |
| Volatiles$_{650-990C}$-wt%: 6.5 | |
| Fixed carbon-wt%: 52.1 | |
| Ash-wt%: 8.2 | |

**Coal, Buelah zap lignite
(North Dakota)**

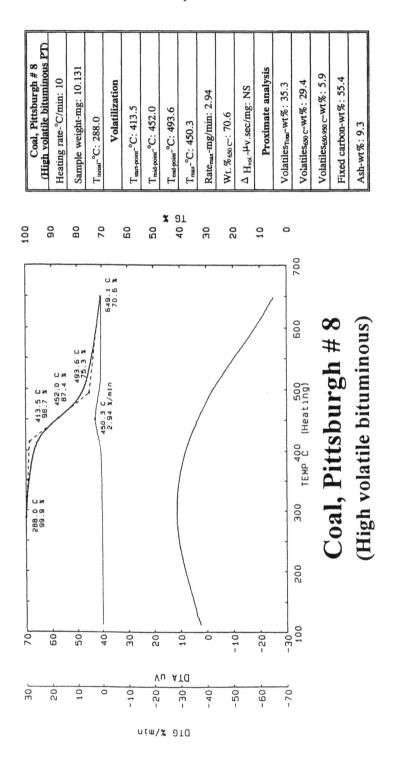

| Coal, Pittsburgh # 8 (High volatile bituminous PT) | |
|---|---|
| Heating rate-°C/min: | 10 |
| Sample weight-mg: | 10.131 |
| $T_{initial}$-°C: | 288.0 |
| **Volatilization** | |
| $T_{start\text{-}point}$-°C: | 413.5 |
| $T_{mid\text{-}point}$-°C: | 452.0 |
| $T_{end\text{-}point}$-°C: | 493.6 |
| $T_{max}$-°C: | 450.3 |
| $Rate_{max}$-mg/min: | 2.94 |
| Wt.%$_{650\,C}$: | 70.6 |
| $\Delta H_{vol.}$-$\mu$v.sec/mg: | NS |
| **Proximate analysis** | |
| Volatiles$_{Total}$-wt%: | 35.3 |
| Volatiles$_{650\,C}$-wt%: | 29.4 |
| Volatiles$_{650\text{-}950\,C}$-wt%: | 5.9 |
| Fixed carbon-wt%: | 55.4 |
| Ash-wt%: | 9.3 |

## Coal, Pittsburgh # 8
### (High volatile bituminous)

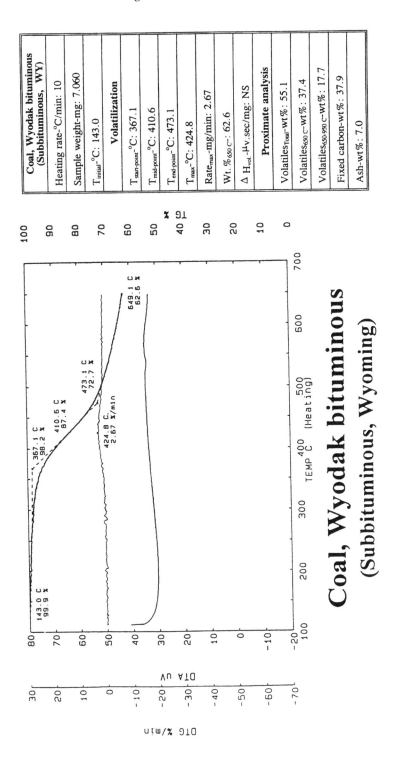

| Coal, Wyodak bituminous (Subbituminous, WY) |
| --- |
| Heating rate-°C/min: 10 |
| Sample weight-mg: 7.060 |
| $T_{initial}$-°C: 143.0 |
| **Volatilization** |
| $T_{start-point}$-°C: 367.1 |
| $T_{mid-point}$-°C: 410.6 |
| $T_{end-point}$-°C: 473.1 |
| $T_{max}$-°C: 424.8 |
| $Rate_{max}$-mg/min: 2.67 |
| Wt.%$_{650C}$: 62.6 |
| $\Delta H_{vol.}$-$\mu$v.sec/mg: NS |
| **Proximate analysis** |
| Volatiles$_{Total}$-wt%: 55.1 |
| Volatiles$_{650C}$-wt%: 37.4 |
| Volatiles$_{650-950C}$-wt%: 17.7 |
| Fixed carbon-wt%: 37.9 |
| Ash-wt%: 7.0 |

# Coal, Wyodak bituminous
## (Subbituminous, Wyoming)

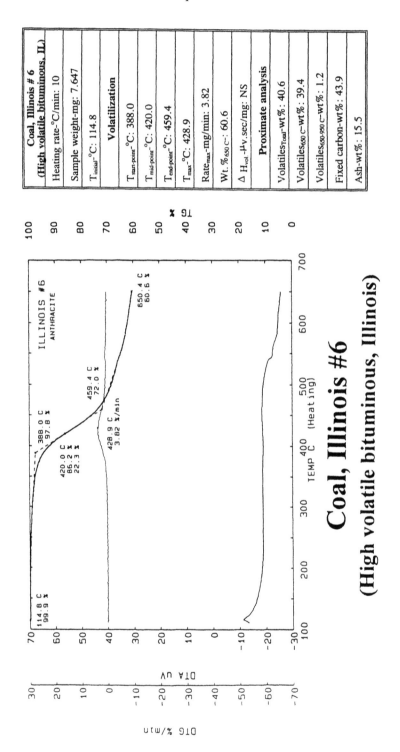

| Coal, Illinois # 6 (High volatile bituminous, IL) | |
|---|---|
| Heating rate-°C/min: | 10 |
| Sample weight-mg: | 7.647 |
| $T_{initial}$-°C: | 114.8 |
| **Volatilization** | |
| $T_{start-point}$-°C: | 388.0 |
| $T_{mid-point}$-°C: | 420.0 |
| $T_{end-point}$-°C: | 459.4 |
| $T_{max}$-°C: | 428.9 |
| $Rate_{max}$-mg/min: | 3.82 |
| Wt. %$_{650 C}$: | 60.6 |
| $\Delta H_{vol}$-μv.sec/mg: | NS |
| **Proximate analysis** | |
| Volatiles$_{T_{total}}$-wt%: | 40.6 |
| Volatiles$_{650 C}$-wt%: | 39.4 |
| Volatiles$_{650-990 C}$-wt%: | 1.2 |
| Fixed carbon-wt%: | 43.9 |
| Ash-wt%: | 15.5 |

# Coal, Illinois #6

## (High volatile bituminous, Illinois)

214

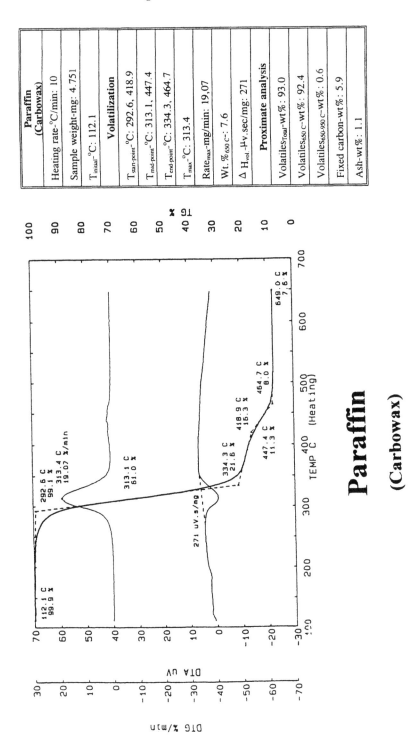

| Paraffin (Carbowax) | |
|---|---|
| Heating rate-°C/min: | 10 |
| Sample weight-mg: | 4.751 |
| $T_{initial}$-°C: | 112.1 |
| **Volatilization** | |
| $T_{start-point}$-°C: | 292.6, 418.9 |
| $T_{mid-point}$-°C: | 313.1, 447.4 |
| $T_{end-point}$-°C: | 334.3, 464.7 |
| $T_{max}$-°C: | 313.4 |
| $Rate_{max}$-mg/min: | 19,07 |
| Wt.$\%_{650C}$: | 7.6 |
| $\Delta H_{vol}$-µv.sec/mg: | 271 |
| **Proximate analysis** | |
| Volatile$_{Total}$-wt%: | 93.0 |
| Volatile$_{S650C}$-wt%: | 92.4 |
| Volatile$_{S650-950C}$-wt%: | 0.6 |
| Fixed carbon-wt%: | 5.9 |
| Ash-wt%: | 1.1 |

# Paraffin
## (Carbowax)

# 11

## LIQUID FUELS

Biomass suffers from being a solid fuel, hard to collect, hard to store, and hard to feed into combustion equipment. The simplest liquid fuels made from biomass are methanol (wood alcohol) from synthesis gas and ethanol (grain alcohol) made by fermentation of sugars from starch or cellulose. However, these chemical compounds have exact boiling points, so that thermogravimetric analysis is not applicable. However, there is wide interest in a number of other liquid fuels from biomass. A recent survey (Stevens, 1994) summarizes ten years of work of the Department of Energy Laboratories on conversion of biomass, primarily to liquid and gas fuels.

### 11.1 VEGETABLE OIL

Vegetable oils such as olive oil and animal fats such as whale oil and tallow have been used for fuels and for cooking for thousands of years. They are remarkably similar in composition, being the esters of glycerol with high-molecular-weight fatty acids of primarily 16–18 carbon atoms. Most animal fats contain saturated fatty acids, primarily stearic ($C_{18}$), and are solid at room temperature. Vegetable-oil fatty acids typically have one or two double bonds (oleic, linoleic acids) and are liquid at room temperature (Swern, 1979).

The very high vaporization temperature of fats and oils (triglycerides) accounts for their use in cooking. However, it makes them less desirable as lighting or diesel fuels.

## 11.2  BIODIESEL

Recently, a superior diesel fuel called biodiesel has been made from the renewable fats and oils by transesterification with methanol or ethanol (Reed et al., 1992; Reed, 1993). The resulting fuel has a high cetane number and low sulfur, which leads to clean combustion in diesel engines.

Conversion of the corn oil to its methyl ester lowers the vaporization temperature by 150°C, which is part of the reason it can be used as an alternate diesel fuel.

## 11.3  PYROLYSIS OILS

If the volatile materials emitted during charcoal manufacture (slow pyrolysis) are condensed, they form an aqueous layer, pyroligneous acid, and an insoluble layer of tar. Many chemicals can be separated from these materials and they formed the basis of our chemical industries (along with coal volatiies) through the 1930s.

When biomass is heated rapidly to about 200°C it begins to decompose into a liquid consisting of the monomer, oligomers, and fragments of the polymers hemicellulose, cellulose, and lignin. These oils have been called "fast pyrolysis" and they can be used as boiler fuels, for chemical synthesis, and even as a diesel fuel. A recent conference explored the use of these oils (NREL, 1994).

Processes are now being developed that give yields of pyrolysis oils up to 70% (including 10% water) (Diebold and Schahill, 1988; Graham et al., 1994; NREL, 1994). A number of laboratories are now producing pyrolysis oils on an experimental basis. Several fast pyrolysis oils are included in this section.

The TG analysis of fast pyrolysis oils from biomass show a very wide temperature range of volatility with 25–30% formation of char. The loss of about 10% water at 120°C is also recorded. Vaporization of the pyrolysis oil from rubber tires occurs below 300°C and produces very little char. The tire oil primarily contains hydrocarbons, particularly limonene.

## 11.4  HYDROCARBONS

Crude oil has provided most of the liquid fuel for the 20th century and will continue to provide energy for the next century until it is gone or environmental considerations limit its use. A thermograms of diesel is included in this atlas for purposes of comparison.

Many plants, such as bayberry, produce a waxy material that is waterproof and protect the outer skin. These waxes have been used for candles. Paraffin wax is refined from paraffinic petroleum oil. While not a liquid fuel at room temperature, its low melting point permits it to be used like liquid fuels in many applications.

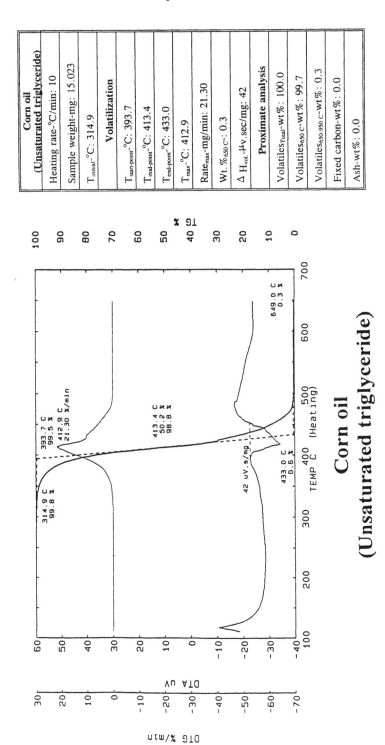

| Corn oil (Unsaturated triglyceride) | |
| --- | --- |
| Heating rate-°C/min: | 10 |
| Sample weight-mg: | 15.023 |
| $T_{initial}$-°C: | 314.9 |
| **Volatilization** | |
| $T_{start-point}$-°C: | 393.7 |
| $T_{mid-point}$-°C: | 413.4 |
| $T_{end-point}$-°C: | 433.0 |
| $T_{max}$-°C: | 412.9 |
| Rate$_{max}$-mg/min: | 21.30 |
| Wt.%$_{650 C}$: | 0.3 |
| $\Delta H_{vol}$-$\mu$v.sec/mg: | 42 |
| **Proximate analysis** | |
| Volatiles$_{Total}$-wt%: | 100.0 |
| Volatiles$_{650 C}$-wt%: | 99.7 |
| Volatiles$_{650-950 C}$-wt%: | 0.3 |
| Fixed carbon-wt%: | 0.0 |
| Ash-wt%: | 0.0 |

**Corn oil
(Unsaturated triglyceride)**

219

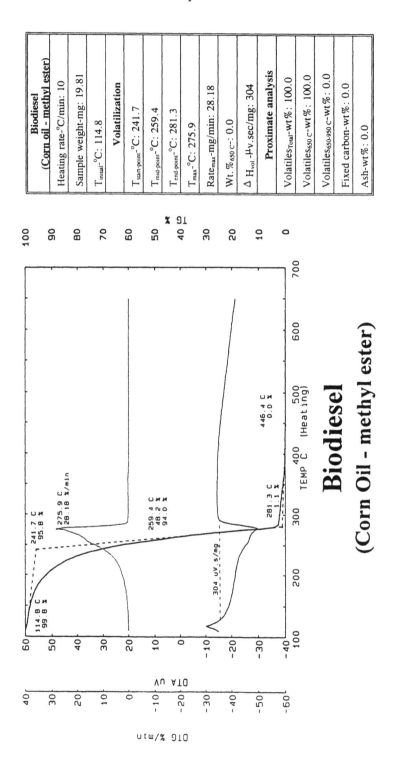

| Biodiesel (Corn oil - methyl ester) | |
|---|---|
| Heating rate-°C/min: 10 | |
| Sample weight-mg: 19.81 | |
| $T_{initial}$-°C: 114.8 | |
| **Volatilization** | |
| $T_{start-point}$-°C: 241.7 | |
| $T_{mid-point}$-°C: 259.4 | |
| $T_{end-point}$-°C: 281.3 | |
| $T_{max}$-°C: 275.9 | |
| $Rate_{max}$-mg/min: 28.18 | |
| Wt.%$_{650\,C}$: 0.0 | |
| $\Delta H_{vol.}$-$\mu$v.sec/mg: 304 | |
| **Proximate analysis** | |
| Volatiles$_{Total}$-wt%: 100.0 | |
| Volatiles$_{550\,C}$-wt%: 100.0 | |
| Volatiles$_{650-950\,C}$-wt%: 0.0 | |
| Fixed carbon-wt%: 0.0 | |
| Ash-wt%: 0.0 | |

## Biodiesel
## (Corn Oil - methyl ester)

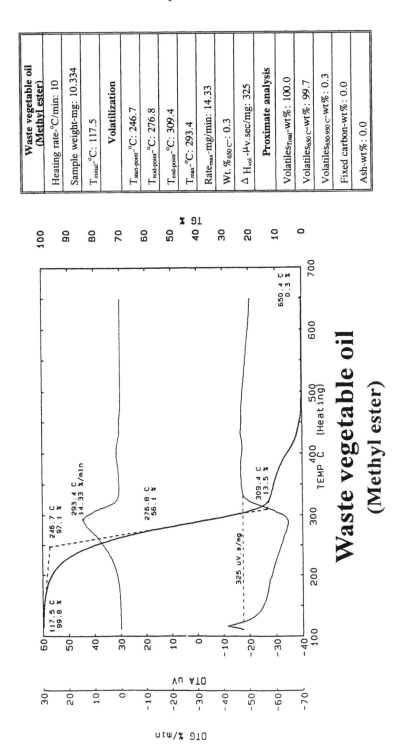

| Waste vegetable oil (Methyl ester) | |
|---|---|
| Heating rate-°C/min: 10 | |
| Sample weight-mg: 10.334 | |
| $T_{initial}$-°C: 117.5 | |
| **Volatilization** | |
| $T_{start-point}$-°C: 246.7 | |
| $T_{mid-point}$-°C: 276.8 | |
| $T_{end-point}$-°C: 309.4 | |
| $T_{max}$-°C: 293.4 | |
| $Rate_{max}$-mg/min: 14.33 | |
| Wt.%$_{650C}$: 0.3 | |
| $\Delta H_{vol.}$-μv.sec/mg: 325 | |
| **Proximate analysis** | |
| Volatiles$_{Total}$-wt%: 100.0 | |
| Volatiles$_{650C}$-wt%: 99.7 | |
| Volatiles$_{650-950C}$-wt%: 0.3 | |
| Fixed carbon-wt%: 0.0 | |
| Ash-wt%: 0.0 | |

## Waste vegetable oil
## (Methyl ester)

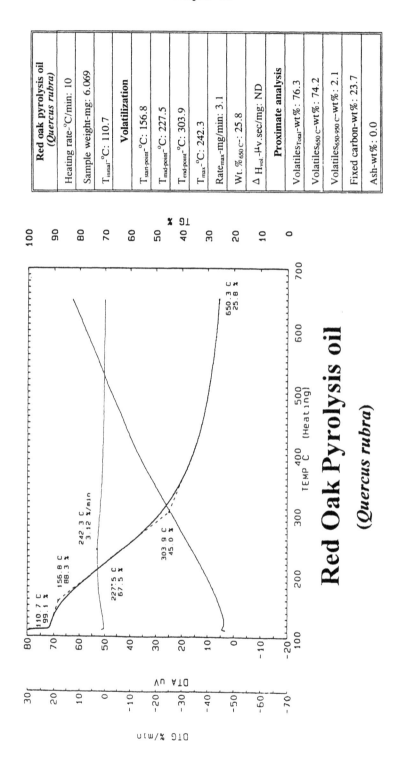

| Red oak pyrolysis oil (*Quercus rubra*) | |
| --- | --- |
| Heating rate-°C/min: | 10 |
| Sample weight-mg: | 6.069 |
| $T_{initial}$-°C: | 110.7 |
| **Volatilization** | |
| $T_{start-point}$-°C: | 156.8 |
| $T_{mid-point}$-°C: | 227.5 |
| $T_{end-point}$-°C: | 303.9 |
| $T_{max}$-°C: | 242.3 |
| $Rate_{max}$-mg/min: | 3.1 |
| Wt.%$_{650\,C}$: | 25.8 |
| $\Delta H_{vol}$-$\mu$v.sec/mg: | ND |
| **Proximate analysis** | |
| Volatiles$_{Total}$-wt%: | 76.3 |
| Volatiles$_{650\,C}$-wt%: | 74.2 |
| Volatiles$_{650-950\,C}$-wt%: | 2.1 |
| Fixed carbon-wt%: | 23.7 |
| Ash-wt%: | 0.0 |

# Red Oak Pyrolysis oil
## (*Quercus rubra*)

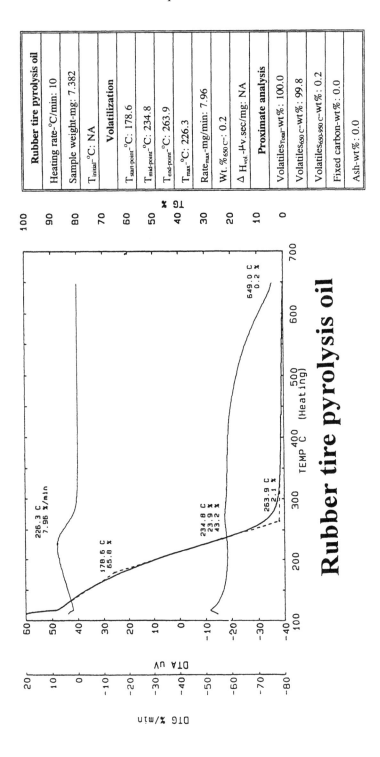

| Rubber tire pyrolysis oil | |
|---|---|
| Heating rate-°C/min: 10 | |
| Sample weight-mg: 7.382 | |
| $T_{initial}$-°C: NA | |
| **Volatilization** | |
| $T_{start-point}$-°C: 178.6 | |
| $T_{mid-point}$-°C: 234.8 | |
| $T_{end-point}$-°C: 263.9 | |
| $T_{max}$-°C: 226.3 | |
| $Rate_{max}$-mg/min: 7.96 | |
| Wt.%$_{650°C}$: 0.2 | |
| $\Delta H_{vol}$-$\mu$v.sec/mg: NA | |
| **Proximate analysis** | |
| Volatiles$_{Total}$-wt%: 100.0 | |
| Volatiles$_{650°C}$-wt%: 99.8 | |
| Volatiles$_{650-950°C}$-wt%: 0.2 | |
| Fixed carbon-wt%: 0.0 | |
| Ash-wt%: 0.0 | |

# Rubber tire pyrolysis oil

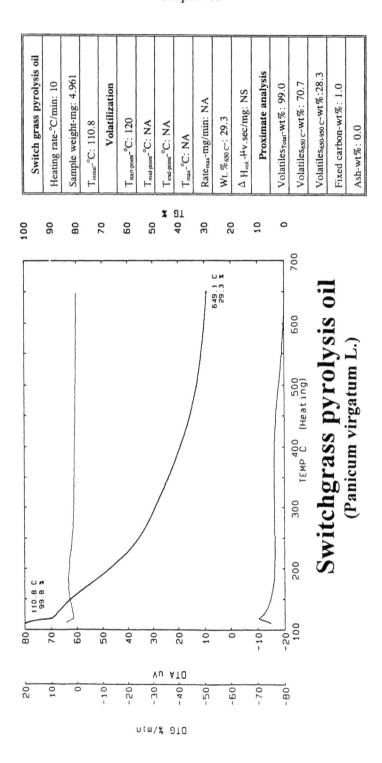

# Switchgrass pyrolysis oil
### (*Panicum virgatum* L.)

| Switch grass pyrolysis oil | |
|---|---|
| Heating rate-°C/min: 10 | |
| Sample weight-mg: 4.961 | |
| $T_{initial}$-°C: 110.8 | |
| **Volatilization** | |
| $T_{start-point}$-°C: 120 | |
| $T_{mid-point}$-°C: NA | |
| $T_{end-point}$-°C: NA | |
| $T_{max}$-°C: NA | |
| $Rate_{max}$-mg/min: NA | |
| Wt.%$_{650C}$-: 29.3 | |
| $\Delta H_{vol}$-$\mu$v.sec/mg: NS | |
| **Proximate analysis** | |
| Volatiles$_{Total}$-wt%: 99.0 | |
| Volatiles$_{650C}$-wt%: 70.7 | |
| Volatiles$_{650-950C}$-wt%:28.3 | |
| Fixed carbon-wt%: 1.0 | |
| Ash-wt%: 0.0 | |

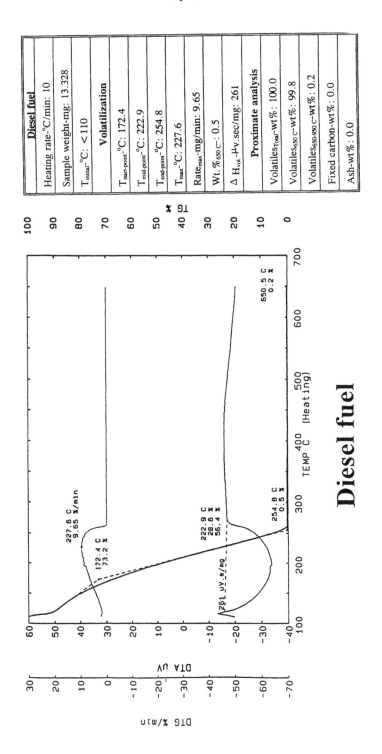

| Diesel fuel | |
|---|---|
| Heating rate-°C/min: | 10 |
| Sample weight-mg: | 13.328 |
| $T_{initial}$-°C: | <110 |
| **Volatilization** | |
| $T_{start\text{-}point}$-°C: | 172.4 |
| $T_{mid\text{-}point}$-°C: | 222.9 |
| $T_{end\text{-}point}$-°C: | 254.8 |
| $T_{max}$-°C: | 227.6 |
| $Rate_{max}$-mg/min: | 9.65 |
| Wt.%$_{650\,C}$: | 0.5 |
| $\Delta H_{vol}$-$\mu$v.sec/mg: | 261 |
| **Proximate analysis** | |
| Volatiles$_{Total}$-wt%: | 100.0 |
| Volatiles$_{650\,C}$-wt%: | 99.8 |
| Volatiles$_{650\text{-}950\,C}$-wt%: | 0.2 |
| Fixed carbon-wt%: | 0.0 |
| Ash-wt%: | 0.0 |

## Diesel fuel

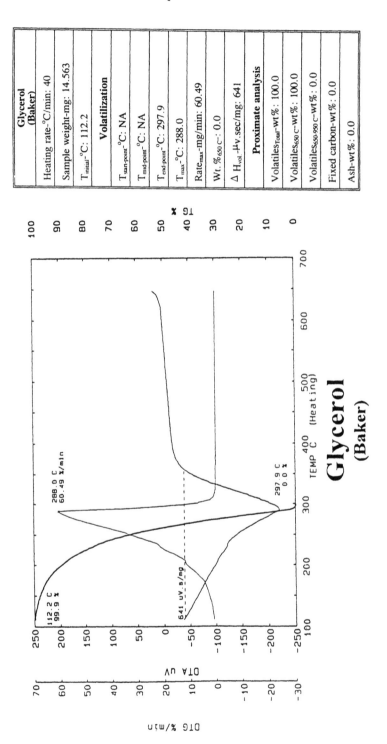

| Glycerol (Baker) | |
|---|---|
| Heating rate-°C/min: | 40 |
| Sample weight-mg: | 14.563 |
| $T_{initial}$-°C: | 112.2 |
| **Volatilization** | |
| $T_{start-point}$-°C: | NA |
| $T_{mid-point}$-°C: | NA |
| $T_{end-point}$-°C: | 297.9 |
| $T_{max}$-°C: | 288.0 |
| $Rate_{max}$-mg/min: | 60.49 |
| Wt.%$_{650 C}$: | 0.0 |
| $\Delta H_{vol}$-μv.sec/mg: | 641 |
| **Proximate analysis** | |
| Volatiles$_{Tend}$-wt%: | 100.0 |
| Volatiles$_{650 C}$-wt%: | 100.0 |
| Volatiles$_{650-950 C}$-wt%: | 0.0 |
| Fixed carbon-wt%: | 0.0 |
| Ash-wt%: | 0.0 |

## Glycerol
## (Baker)

226

# 12

## EFFECT OF OPERATING CONDITIONS

The thermograms shown in this book were taken at one set of conditions except where otherwise noted, and thus are directly comparable. However, the data can be extrapolated to a wider set of conditions. In this chapter, the effects of varying the standard experimental conditions are shown so that the user can estimate the effect of various conditions.

### 12.1   EFFECT OF HEATING RATE

The heating rate is the most important factor in determining the temperature at which the decomposition reactions occur. It has been noted by many researchers that an increase in heating rate causes the thermogram to shift towards higher temperatures, and the effect is predicted theoretically in kinetic studies (see Chapters 1 and 3). It is important to understand this effect because most of the thermal data are collected at low heating rates, while the engineering application of kinetic parameters obtained from this study may be applied to cases where the heating rate is several orders of magnitude higher or lower.

We have provided thermograms for powdered (Western red ceder) and pelletized (Baker cellulose) samples at different heating rates to enable the user to compare their data with the use of kinetic equations given in Chapter 3. In our studies, we found that a tenfold change in heating rate causes a proportional shift in the apparent degradation temperature by about 50–70°C.

## 12.2 EFFECT OF SAMPLE SIZE

The sample size is an important aspect in determining the temperature of decomposition. For large sample size, say greater than 10 –20 mg for most of the biomass materials at heating rates greater than 10°C/min, there is a time lag for the heat transferred from the sample surface to its center, which causes the establishment of a temperature gradient within the sample. This then makes the temperature of the sample measured by the TG less meaningful, because the temperature gradient tends to smear the TG graph to unknown higher temperature. We have shown this effect in a thermogram on a small and large sample. One way to control this effect is to decrease the sample size with the increase in heating rate. This is equivalent to keeping the Biot number of the sample less than 0.1. This aspect is discussed in Chapter 4.

## 12.3 TG ADDITIVITY

It is an unwritten assumption of TG that a mechanical mixture of materials would give a thermogram that would be the arithmetic sum of the proportional quantities of each component. It is not quite so evident that this would occur for the components of biomass that are intimately mixed at the molecular level. It has been shown by Antal (1994) that kinetic measurements on several components can be added arithmetically to give the observed TG curve. Fritsky et al. (1994) have shown that the TG curves for several components of MSW can be added arithmetically to give an observed mixture. Further research related to this aspect is being conducted by the author (Gaur); however, it is appropriate to mention at this stage that initial results agree with the arguments presented by Fritsky et al. (1994).

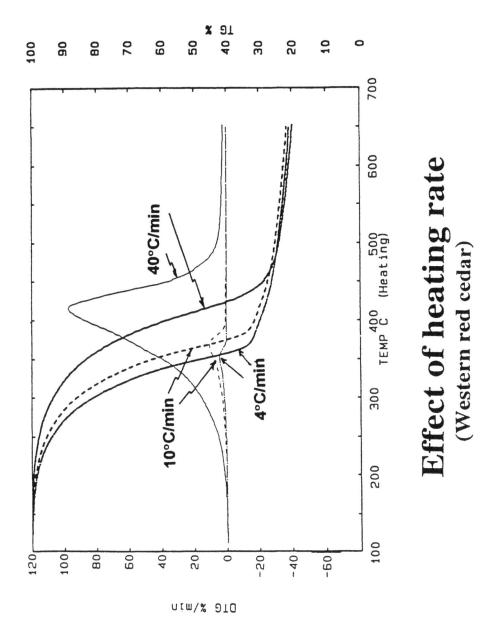

**Effect of heating rate**
(Western red cedar)

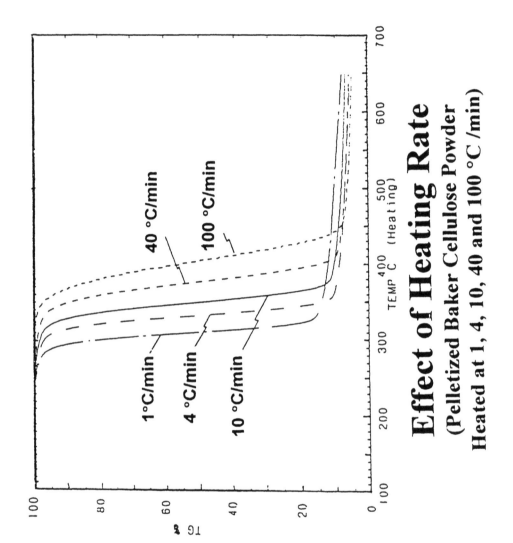

# Effect of Heating Rate
## (Pelletized Baker Cellulose Powder
## Heated at 1, 4, 10, 40 and 100 °C /min)

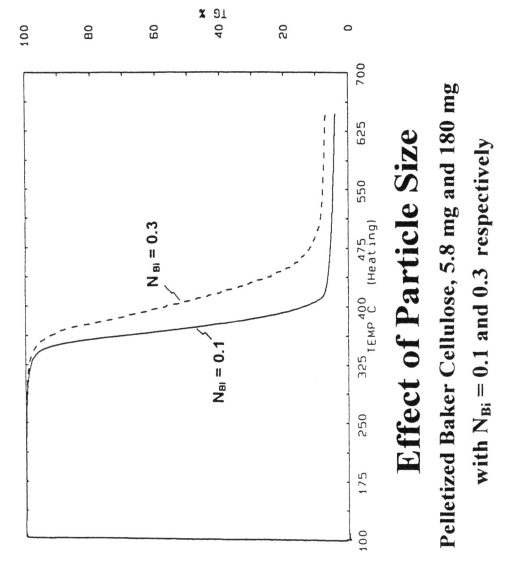

## Effect of Particle Size

**Pelletized Baker Cellulose, 5.8 mg and 180 mg**

**with $N_{Bi} = 0.1$ and 0.3 respectively**

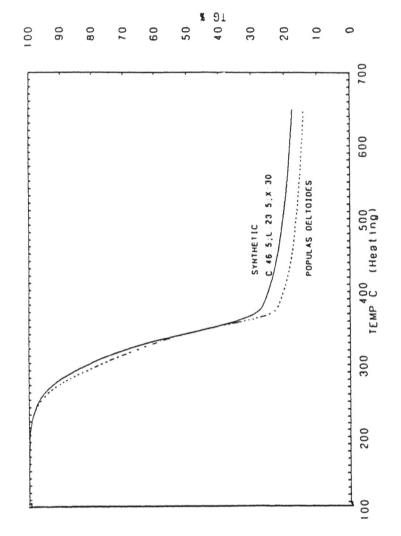

# Biomass Component Additivity

## (*Populus deltoides* - Cell 46.5%; Lignin 23.5%; Xylan 30%)

# APPENDIX A

## PROXIMATE AND ULTIMATE ANALYSIS OF FUELS

### A BRIEF NOTE ON PROXIMATE AND ULTIMATE ANALYSIS

Proximate and ultimate analyses were initially developed for coal, but the same ASTM methods have also been widely used for biomass engineering work. These analyses are usually performed and reported together. The proximate analysis supplies only a minimum amount of data in terms of thermal behavior of any organic sample compared to TGA and DTA. But since it is much easier to conduct proximate and ultimate analysis measurements, a large quantity of such data is available in the literature.

In principle it should also be possible to obtain the proximate analysis from the TGA curves given in this book, since we have determined the amount of volatiles and fixed carbon for all samples. We have also performed proximate analysis measurements using ASTM methods on most of the samples reported in this book, and this analysis is presented in the sidebars for each thermogram. However, there may be some difference in proximate analysis measurements between ASTM and TG techniques. This is primarily because TGA/DTA measurements are made using significantly small sample sizes (1–10 mg) to avoid large heat-transfer resistances due to nonisothermal conditions; in contrast, standard proximate analysis uses relatively large samples, of about 5 g. The fixed carbon content can, therefore, be higher and the volatile content lower in the ASTM proximate analysis measurements when compared with the results obtained from TGA/DTA measurements.

In this appendix, we present the proximate and ultimate analysis data reported by other works on various samples so that the reader will have easy access to information for comparison purposes. However, caution should be exercised when making comparisons with the work presented in this book because most of these analyses were conducted by different researchers using a variety of sample sources and experiment conditions, which could lead to significant deviations.

**TABLE A.1**

Proximate and Ultimate Analysis of Fuels: Biomass

| Name | Fixed carbon (%) | Volatiles (%) | Ash (%) | C (%) | H (%) | O (%) | N (%) | S (%) | HHV KJ/G measured | HHV KJ/G calculated | Error (%) |
|---|---|---|---|---|---|---|---|---|---|---|---|
| **WOOD** | | | | | | | | | | | |
| Beech | — | — | 0.65 | 51.64 | 6.26 | 41.45 | 0.00 | 0.00 | 20.38 | 21.10 | −3.57 |
| Black Locust | 18.26 | 80.94 | 0.80 | 50.73 | 5.71 | 41.93 | 0.57 | 0.01 | 19.71 | 20.08 | −1.87 |
| Canyon Live Oak | 11.30 | 88.20 | 0.50 | 47.84 | 5.80 | 45.76 | 0.07 | 0.01 | 18.98 | 18.79 | 0.99 |
| Chaparall | 18.68 | 75.19 | 6.13 | 46.90 | 5.08 | 40.17 | 0.54 | 0.03 | 18.61 | 18.07 | 2.90 |
| Chinkapin | 12.80 | 86.90 | 0.30 | 49.68 | 5.93 | 44.03 | 0.07 | 0.01 | 19.35 | 19.77 | −2.17 |
| Douglas Fir | 17.70 | 81.50 | 0.80 | 52.30 | 6.30 | 40.50 | 0.10 | 0.00 | 21.05 | 21.48 | −2.01 |
| Douglas Fir | 12.60 | 87.30 | 0.10 | 50.64 | 6.18 | 43.00 | 0.06 | 0.02 | 20.38 | 20.51 | −0.67 |
| Hickory | — | — | 0.73 | 47.67 | 6.49 | 43.11 | 0.00 | 0.00 | 20.17 | 20.51 | −1.72 |
| Madrone | 15.10 | 84.50 | 0.30 | 48.56 | 6.05 | 45.08 | 0.05 | 0.02 | 19.14 | 19.42 | −1.42 |
| Manzanita | 17.89 | 81.29 | 0.82 | 48.18 | 5.94 | 44.68 | 0.17 | 0.02 | 19.30 | 19.18 | 0.62 |
| Maple | — | — | 1.35 | 50.64 | 6.02 | 41.74 | 0.25 | 0.00 | 19.96 | 20.42 | −2.33 |
| Ponerosa Pine | 17.17 | 82.54 | 0.29 | 49.25 | 5.99 | 44.36 | 0.06 | 0.03 | 20.02 | 19.66 | 1.80 |
| Poplar | — | — | 0.65 | 51.64 | 6.26 | 41.45 | 0.00 | 0.00 | 20.75 | 21.10 | −1.71 |
| Red Alder | 12.50 | 87.10 | 0.40 | 49.55 | 6.06 | 43.78 | 0.13 | 0.07 | 19.30 | 19.91 | −3.14 |
| Redwood | 16.10 | 83.50 | 0.40 | 53.50 | 5.90 | 40.30 | 0.10 | 0.00 | 21.03 | 21.45 | −2.02 |
| Tan Oak | 12.40 | 87.10 | 0.50 | 48.34 | 6.12 | 44.99 | 0.03 | 0.03 | 19.12 | 19.43 | −1.60 |
| Western Hemlock | 15.20 | 84.80 | 2.20 | 50.40 | 5.80 | 41.10 | 0.10 | 0.10 | 20.05 | 20.14 | −0.44 |
| Yellow Pine | — | — | 1.31 | 52.60 | 7.00 | 40.10 | 0.00 | 0.00 | 22.30 | 22.44 | −0.61 |
| White Fir | 16.58 | 83.17 | 0.25 | 49.00 | 5.98 | 44.75 | 0.05 | 0.01 | 19.95 | 19.52 | 2.16 |
| White Oak | 17.20 | 81.28 | 1.52 | 49.48 | 5.38 | 43.13 | 0.35 | 0.01 | 19.42 | 19.12 | 1.56 |
| Madrone | 12.00 | 87.80 | 0.20 | 48.94 | 6.03 | 44.75 | 0.05 | 0.02 | 19.51 | 19.56 | −0.25 |
| Tan Oak | 9.20 | 90.60 | 0.20 | 48.67 | 6.03 | 44.99 | 0.06 | 0.04 | 18.93 | 19.44 | −2.69 |

**TABLE A.1** Continued

| Name | Fixed carbon (%) | Volatiles (%) | Ash (%) | C (%) | H (%) | O (%) | N (%) | S (%) | HHV KJ/G measured | HHV KJ/G calculated | Error (%) |
|---|---|---|---|---|---|---|---|---|---|---|---|
| Cedar, White | — | — | 0.40 | 48.80 | 6.40 | 44.40 | — | — | 19.54 | 19.98 | −2.25 |
| Cypress | — | — | 0.40 | 55.00 | 6.50 | 38.10 | — | — | 22.96 | 22.91 | 0.21 |
| Pine, White | — | — | 0.10 | 52.60 | 6.10 | 41.20 | — | — | 20.70 | 21.29 | −2.83 |
| Ash, White | — | — | 0.30 | 49.70 | 6.90 | 43.00 | — | — | 20.75 | 21.03 | −1.34 |
| Elm | — | — | 0.70 | 54.40 | 6.60 | 42.30 | — | — | 20.49 | 20.98 | −2.39 |
| Birch, White | — | — | 0.30 | 49.80 | 6.50 | 43.40 | — | — | 20.12 | 20.55 | −2.14 |
| Mango Wood | 11.36 | 85.64 | 2.98 | 46.24 | 6.08 | 44.42 | 0.28 | — | 19.17 | 18.65 | 2.73 |
| **BARK** | | | | | | | | | | | |
| Douglas Fir Bark | 25.80 | 73.00 | 1.20 | 56.20 | 5.90 | 36.70 | 0.00 | 0.00 | 22.10 | 22.75 | −2.96 |
| Loblolly Pine Bark | 33.90 | 54.70 | 0.40 | 56.30 | 5.60 | 37.70 | 0.00 | 0.00 | 21.78 | 22.35 | −2.37 |
| Long Leaf Pine Bark | 32.30 | 67.00 | 0.70 | 56.40 | 5.50 | 37.40 | 0.00 | 0.00 | 21.77 | 22.29 | −1.56 |
| Slash Pine Bark | 33.40 | 56.90 | 0.70 | 56.20 | 5.40 | 37.30 | 0.00 | 0.00 | 21.77 | 22.11 | −1.56 |
| Cedar | 21.90 | 73.00 | 5.10 | 51.00 | 5.70 | 38.20 | — | — | 20.03 | 20.46 | −2.17 |
| Spruce | 26.60 | 69.60 | 3.80 | 51.80 | 5.70 | 38.60 | — | 0.10 | 20.33 | 20.74 | −2.01 |
| **ENERGY CROPS** | | | | | | | | | | | |
| Eucalyptus Camaldulensis | 17.82 | 81.42 | 0.76 | 49.00 | 5.87 | 43.97 | 0.30 | 0.01 | 19.42 | 19.46 | −0.19 |
| Eucalyptus Globulus | 17.30 | 81.60 | 1.10 | 48.18 | 5.92 | 44.18 | 0.39 | 0.01 | 19.23 | 19.20 | 0.16 |
| Eucalyptus Grandis | 16.93 | 82.55 | 0.52 | 48.33 | 5.89 | 45.13 | 0.15 | 0.01 | 19.35 | 19.13 | 1.12 |
| Casuarina | 19.66 | 78.94 | 1.40 | 48.61 | 5.83 | 43.36 | 0.59 | 0.02 | 19.44 | 19.32 | 0.62 |
| Casuarina | 19.58 | 78.58 | 1.83 | 48.50 | 6.04 | 43.32 | 0.31 | 0.00 | 18.77 | 19.53 | −4.03 |
| Poplar | 16.35 | 82.32 | 1.33 | 48.45 | 5.85 | 43.69 | 0.47 | 0.01 | 19.38 | 19.26 | 0.64 |
| Sudan Grass | 18.60 | 72.75 | 8.65 | 44.58 | 5.35 | 39.18 | 1.21 | 0.01 | 17.39 | 17.62 | −1.30 |

## PROCESSED BIOMASS

| | | | | | | | | | | | |
|---|---|---|---|---|---|---|---|---|---|---|---|
| Plywood | 15.77 | 82.14 | 2.09 | 48.13 | 5.87 | 42.46 | 1.45 | 0.00 | 18.96 | 19.26 | −1.62 |

## AGRICULTURAL

| | | | | | | | | | | | |
|---|---|---|---|---|---|---|---|---|---|---|---|
| Olive Pits | 21.20 | 75.60 | 3.20 | 48.81 | 6.23 | 43.48 | 0.36 | 0.01 | 21.39 | 19.81 | 7.36 |
| Peach Pits | 19.85 | 79.12 | 1.03 | 53.00 | 5.90 | 39.14 | 0.32 | 0.05 | 20.82 | 21.39 | −2.72 |
| Peach Pits | 19.80 | 79.10 | 1.10 | 49.14 | 6.34 | 43.52 | 0.48 | 0.02 | 19.42 | 20.10 | −3.47 |
| Coconut Shells | 22.10 | 77.19 | 0.71 | 50.22 | 5.70 | 43.37 | 0.00 | 0.00 | 20.50 | 19.75 | 3.66 |
| Almond Shells | 21.74 | 73.45 | 4.81 | 44.98 | 5.97 | 42.27 | 1.16 | 0.02 | 19.38 | 18.25 | 5.84 |
| Macadamia Shells | 23.68 | 75.92 | 0.40 | 54.41 | 4.99 | 39.69 | 0.36 | 0.01 | 21.01 | 20.76 | 1.20 |
| Pistachio Shells | 16.84 | 82.03 | 1.13 | 48.79 | 5.91 | 43.41 | 0.56 | 0.01 | 19.26 | 19.48 | −1.12 |
| Walnut Shells | 21.16 | 78.28 | 0.56 | 49.98 | 5.71 | 43.35 | 0.21 | 0.01 | 20.18 | 19.68 | 2.48 |
| Almond Prunings | 21.54 | 76.83 | 1.63 | 51.30 | 5.29 | 40.90 | 0.66 | 0.01 | 20.01 | 19.87 | 0.70 |
| Black Walnut Prunings | 18.56 | 80.69 | 0.78 | 49.80 | 5.82 | 43.25 | 0.22 | 0.01 | 19.83 | 19.75 | 0.39 |
| English Walnut Prunings | 18.10 | 80.82 | 1.08 | 49.72 | 5.63 | 43.14 | 0.37 | 0.01 | 19.63 | 19.50 | 0.65 |
| Cabernet Sauvignon Prunings | 19.20 | 78.63 | 2.17 | 46.59 | 5.85 | 43.90 | 0.83 | 0.04 | 19.03 | 18.56 | 2.45 |
| Chenin Blanc Prunings | 20.21 | 77.28 | 2.51 | 48.02 | 5.89 | 41.93 | 0.86 | 0.07 | 19.13 | 19.31 | −0.94 |
| Corncobs | 18.54 | 80.10 | 1.36 | 46.58 | 5.87 | 45.46 | 0.47 | 0.01 | 18.77 | 18.44 | 1.75 |
| Millet Straw | 16.45 | 78.28 | 5.27 | 43.71 | 5.85 | 45.16 | 0.01 | 0.00 | 18.05 | 17.37 | 3.75 |
| Alfalfa Seed Straw | 20.15 | 72.60 | 7.25 | 46.76 | 5.40 | 40.72 | 1.00 | 0.02 | 18.45 | 18.31 | 0.76 |
| Bean Straw | 18.77 | 75.30 | 5.93 | 42.97 | 5.59 | 44.93 | 0.83 | 0.01 | 17.46 | 16.81 | 3.75 |
| Rice Straw | 17.25 | 69.33 | 13.42 | 41.78 | 4.63 | 36.57 | 0.70 | 0.08 | 16.28 | 15.97 | 1.88 |
| Wheat Straw | 19.80 | 71.30 | 8.90 | 43.20 | 5.00 | 39.40 | 0.61 | 0.11 | 17.51 | 16.71 | 4.55 |
| Corn Stalk | 20.30 | 73.30 | 6.40 | 43.86 | 5.77 | 43.24 | 1.28 | 0.05 | 18.25 | 17.49 | 4.18 |
| Cotton Stalk 1 | 19.90 | 62.90 | 17.20 | 39.47 | 5.07 | 38.09 | 1.25 | 0.02 | 15.83 | 15.44 | 2.50 |
| Cotton Stalk 2 | 22.43 | 70.89 | 6.68 | 43.64 | 5.81 | 43.87 | 0.00 | 0.00 | 18.26 | 17.40 | 4.69 |
| Corn Stover 1 | 19.25 | 75.17 | 5.58 | 43.65 | 5.56 | 43.31 | 0.61 | 0.01 | 17.65 | 17.19 | 2.63 |
| Corn Stover 2 | 14.50 | 78.10 | 7.40 | 46.50 | 5.81 | 39.67 | 0.56 | 0.11 | 19.00 | 18.82 | 0.91 |
| Bagasse | 16.87 | 75.10 | 8.03 | 45.71 | 5.89 | 40.37 | 0.00 | 0.00 | 19.50 | 18.55 | 4.85 |

**TABLE A.1** Continued

| Name | Fixed carbon (%) | Volatiles (%) | Ash (%) | C (%) | H (%) | O (%) | N (%) | S (%) | HHV KJ/G measured | HHV KJ/G calculated | Error (%) |
|---|---|---|---|---|---|---|---|---|---|---|---|
| Bagasse | 22.10 | 64.60 | 13.30 | 39.70 | 5.50 | 42.30 | 0.30 | — | 15.68 | 15.68 | -0.02 |
| Sugarcane Bagasse | 14.95 | 73.78 | 11.27 | 44.80 | 5.35 | 39.55 | 0.38 | 0.01 | 17.33 | 17.61 | -1.63 |
| Coconut Fiber (Coir) | 29.70 | 66.58 | 3.72 | 50.29 | 5.05 | 39.63 | 0.45 | 0.16 | 20.05 | 19.34 | 3.54 |
| Rice Hulls | 15.80 | 63.60 | 20.60 | 38.30 | 4.36 | 35.45 | 0.83 | 0.06 | 14.89 | 14.40 | 3.26 |
| Rice Husk (Patni-23) | 14.90 | 69.30 | 15.80 | 38.92 | 5.10 | 37.89 | 21.7 | 0.12 | 15.67 | 15.32 | 2.23 |
| Rice Husk Bran | 19.53 | 61.83 | 18.64 | 38.92 | 5.12 | 36.77 | 0.55 | 0.00 | 15.29 | 15.42 | -0.82 |
| Almond Hulls | 22.89 | 71.33 | 5.78 | 45.79 | 5.36 | 40.60 | 0.96 | 0.01 | 18.22 | 17.97 | 1.39 |
| Cocoa Hulls | 23.80 | 67.95 | 8.25 | 48.23 | 5.23 | 33.19 | 2.98 | 0.12 | 19.04 | 19.36 | -1.69 |
| Peanut Hulls | 21.09 | 73.02 | 5.89 | 45.77 | 5.46 | 39.56 | 1.63 | 0.12 | 18.64 | 18.18 | 2.45 |
| Groundnut Shells | 21.60 | 72.70 | 5.70 | 48.59 | 5.64 | 39.49 | 0.58 | — | 19.85 | 19.40 | 2.28 |
| Wheat Dust | 16.47 | 69.85 | 13.68 | 41.38 | 5.10 | 35.17 | 3.04 | 0.19 | 16.20 | 16.50 | -1.86 |
| Pine Needles | 26.12 | 72.38 | 1.50 | 48.21 | 6.57 | 43.72 | — | — | 20.12 | 20.02 | 0.50 |
| Cotton Gin Waste | 14.97 | 83.41 | 1.61 | 42.66 | 6.05 | 49.50 | 0.18 | — | 17.48 | 16.87 | 3.53 |
| Cotton Gin Trash | 15.10 | 67.30 | 17.60 | 39.59 | 5.26 | 36.38 | 2.09 | 0.00 | 16.42 | 15.85 | 3.45 |
| Grape Pomace | 21.40 | 74.40 | 4.20 | 54.94 | 5.83 | 32.08 | 2.09 | 0.21 | 21.80 | 22.63 | -3.81 |
| Napier Grass | — | — | 5.70 | 45.20 | 6.00 | 42.30 | 0.00 | — | 18.31 | 18.35 | -0.26 |
| **AQUATIC BIOMASS** | | | | | | | | | | | |
| Water Hyacinth (Florida) | — | 80.40 | 19.60 | 40.30 | 4.60 | 33.99 | 1.51 | 0.00 | 14.86 | 15.54 | -4.56 |
| Water Hyacinth | — | — | 15.30 | 43.00 | 5.80 | 29.50 | 5.60 | — | 17.98 | 18.39 | -2.26 |
| Brown Kelp, Giant, Monterey | — | 54.20 | 45.80 | 26.60 | 3.74 | 20.22 | 2.55 | 1.09 | 10.26 | 10.71 | -4.38 |
| Brown Kelp, Giant, Soquel Point | — | 57.90 | 42.10 | 27.80 | 3.77 | 23.69 | 4.63 | 1.05 | 10.75 | 10.85 | -0.94 |

Main source for data on biomass samples: Channiwala, Salim A., On Biomass Gasification Process and Technology Development, Ph.D. Thesis, Indian Institute of Technology, Bombay, India, 1992.

238

**TABLE A.2**

Proximate and Ultimate Analysis of Fuels: Liquid Fuels

| Name | Fixed carbon (%) | Volatiles (%) | Ash (%) | C (%) | H (%) | O (%) | N (%) | S (%) | HHV KJ/G measured | HHV KJ/G calculated | Error (%) |
|---|---|---|---|---|---|---|---|---|---|---|---|
| **HYDROCARBONS** | | | | | | | | | | | |
| Methane, $CH_4$ | 0.00 | 100 | 0.00 | 74.85 | 25.15 | 0.00 | 0.00 | 0.00 | 55.35 | 55.76 | −0.76 |
| n-Octane | 0.00 | 100 | 0.00 | 84.10 | 15.90 | 0.00 | 0.00 | 0.00 | 47.80 | 48.09 | −0.63 |
| Benzene, $C_6H_6$ | 0.00 | 100 | — | 92.25 | 7.75 | 0.00 | 0.00 | 0.00 | 41.79 | 41.34 | 1.08 |
| Motor Gasoline | 0.00 | 100 | — | 85.50 | 14.40 | 0.00 | 0.00 | 0.10 | 46.88 | 46.83 | 0.12 |
| Kerosene | 0.00 | 100 | 0.01 | 85.80 | 14.10 | 0.00 | 0.00 | 0.10 | 46.50 | 46.58 | −0.17 |
| Diesel Oil | 0.00 | 100 | — | 86.50 | 13.20 | 0.00 | 0.00 | 0.30 | 45.70 | 45.78 | −0.18 |
| No. 6 Fuel Oil | 0.00 | 100 | — | 85.70 | 10.50 | 1.70 | 2.00 | 0.05 | 42.30 | 42.09 | 0.50 |
| **ALCOHOLS** | | | | | | | | | | | |
| Methanol, $CH_3OH$ | 0.00 | 100 | 0.00 | 37.50 | 12.50 | 50.00 | 0.00 | 0.00 | 22.69 | 22.65 | 0.18 |
| Ethanol, $C_2H_5OH$ | 0.00 | 100 | 0.00 | 52.20 | 13.00 | 34.80 | 0.00 | 0.00 | 30.15 | 29.94 | 0.69 |
| **PYROLYSIS OILS** | | | | | | | | | | | |
| LBL Wood Oil | — | — | 0.78 | 72.30 | 8.60 | 18.60 | 0.20 | 0.01 | 33.70 | 33.54 | 0.49 |
| BOM Wood Oil | — | — | 0.66 | 82.00 | 8.80 | 9.20 | 0.60 | 0.00 | 36.80 | 38.02 | −3.32 |
| Coke-Oven Tar | — | — | 0.25 | 91.75 | 5.50 | 0.80 | 0.90 | 0.80 | 38.20 | 38.49 | −0.76 |
| Low-Temperature Tar | — | — | — | 83.00 | 8.20 | 7.40 | 0.60 | 0.80 | 38.75 | 37.94 | 2.08 |

**TABLE A.3**

Proximate and Ultimate Analysis of Fuels: Solid Fuels

| Name | Fixed carbon (%) | Volatiles (%) | Ash (%) | C (%) | H (%) | O (%) | N (%) | S (%) | HHV KJ/G measured | HHV KJ/G calculated | Error (%) |
|---|---|---|---|---|---|---|---|---|---|---|---|
| **COAL** | | | | | | | | | | | |
| Pittsburgh Steam | 55.80 | 33.90 | 10.30 | 75.50 | 5.00 | 4.90 | 1.20 | 3.10 | 31.75 | 31.82 | −0.21 |
| Wyoming Elkol | 46.60 | 43.00 | 4.20 | 71.50 | 5.30 | 16.90 | 1.20 | 0.90 | 29.57 | 29.44 | 0.42 |
| Northumberland No. 8 Anthr | 84.59 | 7309.00 | 8.32 | 83.67 | 3.56 | 2.84 | 0.55 | 1.05 | 32.86 | 33.03 | −0.54 |
| German-Anna | 79.60 | 12.00 | 8.40 | 82.62 | 3.02 | 3.66 | 0.92 | 0.73 | 33.00 | 31.91 | 3.32 |
| Green In. #3 | 41.53 | 40.93 | 17.55 | 62.70 | 4.84 | 6.29 | 1.36 | 7.17 | 27.36 | 27.27 | 0.31 |
| German Braunkole Lign | 46.03 | 49.47 | 4.50 | 63.89 | 4.97 | 24.54 | 0.57 | 0.48 | 25.10 | 25.57 | −1.86 |
| Peat, S-H3 | 26.87 | 70.13 | 3.00 | 54.81 | 5.38 | 35.81 | 0.89 | 0.11 | 22.00 | 21.71 | 1.34 |
| **COKE** | | | | | | | | | | | |
| Coke | 91.47 | 0.92 | 7.61 | 89.13 | 0.43 | 0.98 | 0.85 | 1.00 | 31.12 | 31.45 | −1.04 |
| Low-Temperature Coke | 81.79 | 11.15 | 7.06 | 83.53 | 2.82 | 3.76 | 1.00 | 1.82 | 30.94 | 32.11 | −3.79 |
| **CHARCOAL** | | | | | | | | | | | |
| Charcoal | 89.31 | 93.88 | 1.02 | 92.04 | 2.45 | 2.96 | 0.53 | 1.00 | 34.39 | 34.78 | −1.15 |
| Subabul Wood Char (950°C) | 74.50 | 21.65 | 3.95 | 83.61 | 1.95 | 10.48 | 0.01 | 0.00 | 30.35 | 30.32 | 0.10 |
| Redwood Char (550°C) | 67.70 | 30.00 | 2.30 | 75.60 | 3.30 | 18.40 | 0.20 | 0.20 | 28.84 | 28.35 | 1.73 |
| Redwood Char (935°C) | 72.00 | 23.90 | 4.10 | 78.80 | 3.50 | 13.20 | 0.20 | 0.20 | 30.47 | 30.20 | 0.90 |
| Oak Char (630°C) | 59.30 | 25.80 | 14.90 | 67.70 | 2.40 | 14.40 | 0.40 | 0.20 | 24.80 | 24.67 | 0.50 |
| Oak Char (565°C) | 55.60 | 27.10 | 17.30 | 64.60 | 2.10 | 15.50 | 0.40 | 0.10 | 23.05 | 23.06 | −0.05 |
| Casuarina Char (950°C) | 71.53 | 15.23 | 13.24 | 77.54 | 0.93 | 5.62 | 2.67 | 0.00 | 27.12 | 27.26 | −0.53 |
| Coconut Shell Char (750°C) | 87.17 | 9393.00 | 2.90 | 88.95 | 0.73 | 6.04 | 1.38 | 0.00 | 31.12 | 31.21 | −0.26 |
| Eucalyptus Char (950°C) | 70.32 | 19.22 | 10.45 | 76.10 | 1.33 | 11.10 | 1.02 | 0.00 | 27.60 | 26.75 | 3.07 |

**TABLE A.4**
Proximate and Ultimate Analysis of Fuels: Exceptions

| Name | Fixed carbon (%) | Volatiles (%) | Ash (%) | C (%) | H (%) | O (%) | N (%) | S (%) | HHV KJ/G measured | HHV KJ/G calculated | Error (%) |
|---|---|---|---|---|---|---|---|---|---|---|---|
| Hydrogen, $H_2$ | — | — | — | 0.00 | 100.0 | 0.00 | 0.00 | 0.00 | 141.26 | 117.83 | 16.59 |
| Carbon Monoxide, CO | — | — | — | 42.86 | 0.00 | 57.14 | 0.00 | 0.00 | 10.16 | 9.05 | 10.89 |
| Acetylene | — | — | — | 92.25 | 7.75 | 0.00 | 0.00 | 0.00 | 49.60 | 41.34 | 16.66 |
| Carbon | — | — | — | 100.00 | — | — | — | — | 32.81 | 34.91 | −6.40 |
| Carbon Dioxide, $CO_2$ | — | — | — | 27.27 | — | — | — | — | 0.00 | 9.52 | NA |
| Water | — | — | — | 0.00 | 11.11 | 88.89 | — | — | 0.00 | 3.90 | NA |
| **NIST SAMPLES** | | | | | | | | | | | |
| Eastern Cottonwood | 15.80 | 77.80 | 0.90 | 46.60 | 6.00 | 42.80 | 0.09 | 0.05 | 18.34 | 18.39 | −3.04 |
| Monterey Pine | 14.60 | 79.90 | 0.30 | 47.10 | 6.10 | 42.60 | 0.05 | 0.01 | 18.22 | 19.22 | −5.51 |
| Wheat Straw | 18.80 | 67.70 | 8.40 | 40.90 | 5.60 | 40.60 | 0.57 | 0.15 | 16.39 | 16.51 | −0.74 |
| Sugar Cane Bagasse | 13.70 | 77.60 | 3.60 | 45.10 | 6.97 | 42.10 | 0.15 | 0.03 | 17.92 | 19.53 | −8.96 |
| **NIST—DRY BASIS** | | | | | | | | | | | |
| Eastern Cottonwood | 16.77 | 82.59 | 0.96 | 49.47 | 6.37 | 45.44 | 0.10 | 0.05 | 19.47 | 20.06 | −3.04 |
| Monterey Pine | 15.55 | 85.09 | 0.32 | 50.16 | 6.50 | 45.37 | 0.05 | 0.01 | 19.40 | 20.47 | −5.51 |
| Wheat Straw | 20.22 | 72.80 | 9.03 | 43.98 | 6.02 | 43.66 | 0.61 | 0.16 | 17.62 | 17.75 | −0.74 |
| Sugar Cane Bagasse | 14.65 | 82.99 | 3.85 | 48.24 | 7.45 | 45.03 | 0.16 | 0.03 | 19.17 | 20.89 | −8.96 |

# APPENDIX B

## Sources of Samples

| Sample Name | Source |
|---|---|
| 25% Cotton mix paper | IPST |
| Acetylated xylan | NREL |
| Avicel PH 102 | CSM |
| Babool wood | CSM |
| Baggase (stored for zero weeks) | NREL |
| Baggase (stored for 26 weeks) | NREL |
| Biodiesel | CSM |
| Biomass components additivity (*populous deltoides*) | NREL |
| Black locust | CSU |
| Blind canyon coal | ANL |
| Brown rotten wood | CSM |
| Buelah zap lignite (North Dakota) | ANL |
| Carbowax | CSM |
| Cellulose | CSM |
| Cellulose (Baker analyzed) | CSM |
| Cellulose (Doped with 0.1% KOH) | NREL |
| Cellulose (Doped with 0.1% $ZnCl_2$) | NREL |
| Cellulose (Synthetic) | CSM |
| Coconut shell | CSM |
| Corn oil | CSM |
| Corn oil ester | CSM |
| Corrugated sheet cardboard | IPST |

| Sample Name | Source |
| --- | --- |
| Cotton (dryland) | NREL |
| Cotton (irrigated) | NREL |
| Diesel fuel | CSM |
| Eastern red cedar | CSU |
| Ecofuel | NREL |
| Eelgrass | MRI |
| Giant kelp | MRI |
| Glucose | CSM |
| Glycerin | CSM |
| Hemlock wood | CSU |
| Illinois #6 coal | ANL |
| Kimwipe tissue paper | CSM |
| Kraft pine lignin | NREL |
| Lactic acid | CSM |
| Lignin | NREL |
| Lignin (*populous deltoides*) | NREL |
| Lignin-aspen (steam exploded) | NREL |
| Liner board | IPST |
| Lodgepole pine | CSU |
| Newsprint paper | IPST |
| Oak bark | CSU |
| Oak wood (pyrolysis oil) | NREL |
| Oil shale | CSM |
| Osage orange | CSU |
| Peach seed | CSM |
| Peanut shell | CSM |
| Peat | IHI |
| Peat (steam exploded) | IHI |
| Pinus radiata | CSU |
| Pistachio nut | CSM |
| Pittsburgh #8 coal | ANL |
| Plant anatomy samples | NREL |
| Plant extract samples | NREL, CSM |
| Plant maturation samples | NREL |
| Polydimethyl siloxane | CSM |
| Polyethylene | CSM |
| Polyethylene glycol | CSM |
| Polyhydroxy benzoic acid | CSM |
| Polymethyl methacrylate | CSM |
| Polystyrene | CSM |
| Polytetrafluoroethylene (Teflon) | CSM |
| Polyvinyl acetate | CSM |
| Polyvinyl chloride | CSM |
| Ponderosa pine | CSU |
| Ponderosa pine bark | CSU |
| Poplar (stored for zero weeks) | NREL |

| Sample Name | Source |
|---|---|
| Poplar (stored for 26 weeks) | NREL |
| Populus deltoides | NREL |
| Populus tremuloides | NREL |
| Prima rayonier | NREL |
| Pumpkin seed | CSM |
| Red alder bark | CSM |
| Refuse-derived fuel (Teledyne) | NREL |
| Refuse-derived fuel (Thief River Falls) | NREL |
| Rice hull | CSM |
| Rubber tire (pyrolysis oil) | NREL |
| Rubber wood | NREL |
| Sargassum weed | MRI |
| Sericea lespedeza | NREL |
| Sericea (stored for zero weeks) | NREL |
| Sericea (stored for 26 weeks) | NREL |
| Slash pine | CSU |
| Sorghum (stored for zero weeks) | NREL |
| Sorghum (stored for 26 weeks) | NREL |
| Sugar pine | CSU |
| Sunflower stalk | CSM |
| Switch grass (pyrolysis oil) | NREL |
| Switch grass (stored for zero weeks) | NREL |
| Switch grass (stored for 26 weeks) | NREL |
| Switch grass char | NREL |
| Walnut | CSM |
| Waste vegetable oil ester | CSM |
| Western red cedar | CSU |
| Wheat straw | CSM |
| White rotten wood | CSM |
| Wyodak bituminous coal | ANL |

ANL, Argonne National Laboratory, Argonne, Illinois.
CSU, Colorado State University, Fort Collins, Colorado; contact person: Dr. H. Schroder.
CSM, Colorado School of Mines, Golden, Colorado; contact person: Drs. T. B. Reed, S. Gaur, and K. Vorhees.
IPST, Institute of Paper Science and Technology, Atlanta, Georgia; contact person: A. W. Rudie.
NREL, National Renewable Energy Laboratory, Golden, Colorado; contact persons: Drs. F. Agblevaor, R. Evans, D. Johnson, and T. A. Milne.
MERI, Marine Research Institute, Brooklin, Maine.
IHI, International Humic Institute, CSM, Golden, Colorado; contact person: Dr. P. McCarthy.

# APPENDIX C

## REFERENCES AND SUGGESTED READINGS

The information presented in this book is filtered from various sources. Some of the aspects that were not directly relevant to the scope of this book have not been quoted in the text. However, it is our view that it would be in the interest of the reader to be aware of various literature sources that are connected with the topics presented in this book. Hence, we have attempted to present the reference list in a slightly more expanded form and hope this aspect is useful to the reader.

Abatzoglou, N., Chornet, E., Belkacemi, K., and Overend, R. P. (1992). Chem. Engrg. Sci., Vol. 47, p. 1109.

Agrawal, R. K. (1986). J. Therm. Analy., Vol. 31, p. 1253.

Akita, K., and Kase, M. (1967). J. Polymeric Sci., A5, p. 833.

Alter, H. (1983). Municipal Solid Waste, Marcel Dekker, Inc., New York.

Antal, M. J., Jr. (1982). Advances in Solar Energy, K. W. Boer and J. A. Duffie, eds., Vol. 1, American Solar Energy Society, Boulder, CO, pp. 61–112.

Antal, M. J., Jr. (1985). Advances in Solar Energy, K. W. Boer and J. A. Duffie, eds., Vol. 1, American Solar Energy Society, Boulder, CO, pp. 175–255.

Antal, M. J., Jr. (1982). "A Review of the Gas Phase Pyrolysis of Biomass Derived Volatile Matter," presented at Fundamentals of Thermal Conversion: An International Conference, Estes Park, CO.

Antal, M. J., Jr., Friedman, H. L., and Rogers, F. E. (1980). Combustion Sci. and Technol., Vol. 21, p. 141.

Antal, M. J., Jr., and Varhegyi, G. (1995). Cellulose pyrolysis kinetics: the current state of knowledge, Ind. Eng. Chem. Res.

Arseneau, D. F. (1971). Competitive reactions in the thermal decomposition of cellulose. Can. J. Chim., Vol. 49, p. 632.

ASTM Standards: Standard Method of Preparing Coal Samples for Analysis, 3172-73; Standard Method for Proximate Analysis of Coal and Coke, 3173-87; Standard Test Method for Moisture in the Analysis of Samples of Coal and Coke, 3175-89; Standard Test Method for Volatile Matter in the Analysis of Samples of Coal and Coke, 3174-89; Standard Test Method for Ash in the Analysis of Samples of Coal and Coke, 3176-84; Standard Method for Ultimate Analysis of Coal and Coke, 3177-84; Standard Test Method for Total Sulfur in the Analysis of Samples of Coal and Coke, 3178-84; Standard Test Methods for Carbon and Hydrogen in the Analysis of Samples of Coals and Cokes, 3179-84; Standard Test Methods for Nitrogen in the Analysis of Samples of Coal and Coke, 2015-85; Standard Test Method for Gross Calorific Value of Coal and Coke by the Adiabatic Bomb Calorimeter, in Coal and Coke, Section 5, Vol. 05.05, Annual Book of ASTM Standards (1989).

Bain, R. (1981). Biomass Gasification: Principles and Technology, T. B. Reed, ed., Noyes Data Corporation, Park Ridge, NJ.

Baker, R. R. (1975). Thermal decomposition of cellulose, J. Thermal Anal., Vol. 8, p. 163.

Barooah, J. N., and Long, V. D. (1976). Fuel, Vol. 55, p. 116.

Basch, A., and Lewin, M. (1973). J. Polym. Sci., Vol. 11, p. 2071.

Bateman, H., and Erdelyi, A. (1953). Higher Transcendental Functions, Vol. 2, McGraw Hill Book Co., New York, NY.

Belkacemi, K., Abatzoglou, N., Overend, R. P., and Chornet, E. (1991). I&EC Research, Vol. 30, p. 2416.

Berger, I. A., and Whitehead, W. L. (1951). Fuel, Vol. 30, p. 247.

Berkowitz-Mattuck, J. B., and Noguchi, T. (1963). Pyrolysis of untreated and APO-THPC treated cotton cellulose during one second exposure to radiant flux levels of 5-25 cal/CM2-SCC, J. Appl. Polym. Sci., 7, 709.

Blumberg, A. J. (1959). J. Phys. Chem., Vol. 63, p. 1129.

Borchardt, H. J., and Daniels, F. J. (1957). Am. Chem. Soc., Vol. 79, p. 41.

Borchardt, H. J. (1956). Ph.D. Thesis, University of Wisconsin, Madison, WI.

Bradbury, A. G. W., Sakai, Y., and Shafizadeh, F. (1979). J. Appl. Polym. Sci., Vol. 23, pp. 3271–3280.

Bridgwater, A. V., ed. (1994). Advances in Thermochemical Biomass Conversion, Blackie Academic and Professional, London, England.

Broido, A., and Kilzer, F. J. (1963). Fire Res. Abs. Rev., Vol. 5, p. 157.

Broido, A., and Weinstein, M. (1970). Comb. Sci. Tech., Vol. 1, pp. 279–285.

Broido, A., and Nelson, M. A. (1975). Char yield on pyrolysis of cellulose. Combust. and Flame, Vol. 24, p. 263.

Broido, A. (1969). J. Polym. Sci., Part A2, Vol. 7, p. 1761.

Broido, A. (1976). Thermal Uses and Properties of Carbohydrates and Lignins. Shafizadeh, F., Sarkanen, K., and Tillman, D., eds., Academic Press, New York, NY, p. 19.

Broido, A. (1966). Thermogravimetric and differential thermal analysis of potassium bicarbonate contaminated cellulose. Pyrodynamics, Vol. 4, p. 243.

Bryce, D. J., and Greenwood, C. T. (1963). Staerke, Vol. 15, pp. 285–290.

Bryce, D. J., and Greenwood, C. T. (1966). Appl. Polym. Symp., Vol 2, p. 149.

Bryne, A., Gardiner, D., and Holmes, F. H. (1966). J. Appl. Chem., Vol. 16, pp. 81–88.

Cardwell, R. D., and Lunar, P. (1976). Wood Sci. Tech., Vol. 10, pp. 131–147.

Cemy, M., and Stanek, J., Jr. (1977). Adv. Carbohyd. Chem., Vol. 34, p. 23.

Chan, R. W., and Krieger, B. B. (1981). J. Appl. Polym. Sci., Vol. 26, p. 1533.

Channiwala, S. A. (1992). Ph.D. Thesis, Indian Institute of Technology, Department of Mechanical Engineering, Bombay, India.

Chatterjee, P. K., and Conrad, C. M. (1966). Kinetics of the pyrolysis of cotton cellulose. Textile Res. J., Vol. 36, p. 487.

Chatterjee, P. K. (1968). Chain reaction mechanism of cellulose pyrolysis. J. Appl. Polym. Sci., Vol. 12, p. 1859.

Chatterjee, P. K. (1965). J. Polym. Sci., Vol. A3, p. 4253.

Chiu, J. (1962). Anal. Chem., Vol. 34, No. 13, p. 1841.

Chiu, J. (1966). Thermoanalysis of Fiber and Fiber Forming Polymers, R. F. Schwenker, ed., Interscience, New York, NY, p. 25.

Chornet, E., and Roy, C. (1980). Thermochemica Acta, Vol. 35, pp. 389–393.

Coats, A. W., and Redfern, J. P. (1964). Nature, Vol. 201, p. 68.

Diebold, J. P. (1981). "Entrained flow ablative fast pyrolysis," 13th Biomass Thermochemical Conversion Contractors Meeting, Arlington, VA.

Diebold, J. P., and Schahill, J. W. (1988). Energy Progress, Vol. 8, No. 1, p. 59.

Diebold, J. P. (1993). "A seven step global model for the pyrolysis of cellulose," National American Chemical Society Meeting, Cellulose and Textile Div., Abstract No. 1, American Chemical Society, Washington, DC.

Diebold, J. P. (1984). "Ablative entrained flow fast pyrolysis of biomass," 16th Biomass Thermochemical Conversion Contractors Meeting, Portland OR.

Diebold, J. P. (1985). M.S. Thesis T 3007, Colorado School of Mines, Golden, CO.

Domburgs, G., Rossinskaya, G., and Sergeeva, N. V. (1974). 4th Therm. Anal. Proc., Vol 2, p. 211.

Doyle (1961). J. Appl. Polym. Sci., Vol. V, No. 15, p. 285.

Dutta, S., Wen, C. Y., and Belt, R. J. (1977). "Reactivity of coal and char," Ind. Eng. Chem. Proc. Des. Dev., Vol. 16, pp. 20–37.

Duval, C. (1951). Anal. Chem., Vol. 23, p. 1271.

Earl, D. E. (1974). A Report on Charcoal, FAO of UN, ISBN 92-5-10024-X.

Ergun, S. (1956). J. Phys. Chem., Vol. 60, p. 480.

Eventova, I. L., et al. (1974). Khim. Volokra, Vol. 4, pp. 29–31.

Eysing, H., and Polanyi, M. (1931). Z. Physik. Chem. B, Vol. 12, No. 279.

Fenner, R. A., and Lephardt, I. D. (1981). J. Agrk. Food Chem., Vol. 29, p. 646.

Flynn, J. H., and Wall, L. A. (1966). J. Res. NBS, Vol. 70A, No. 6, p. 487.

Formella, K., Leonhardt, P. Sulimma, A., van Heek, K. H., and Juntgen, H. (1986). Interaction of mineral matter in coal with potassium during gasification, Fuel, Vol. 65, pp. 1470–1472.

Freeman, D., and Carroll (1958). J. Chem. Phys., Vol. 62, p. 394.

Friedman, H. L. (1965). Analy. Chem., Vol. 37, p. 768.

Friedman, H. L. (1964). Kinetics of thermal degradation of char-forming plastics from thermogravimetry, J. Polym. Sci., Vol. C6, p. 183.

Fritsky, K. J., Miller, D. L., and Cernansky, N. P. (1994). J. Air and Waste Mgmnt. Assn., Vol. 44, p. 1116.

Fung, D. P. C. (1969). Tappi, Vol. 62, p. 319.

Furneaux, R. H., and Shafizadeh, F. (1979). Carbohydrate Res., Vol. 74, p. 354.

Gadalla, A. M. (1985). Thermochimica Acta, Vol. 95, p. 179.

Garn, P. D., and Flashen, S. S. (1957). Anal. Chem., Vol. 29, p. 271.

Gaur, S., and Reed, T. B. (1994). Biomass and Bioenergy, Vol. 7, 1–6, 61–67.

Gaur, S. (1989). Ph.D. Thesis, Indian Institute of Technology, New Delhi, India.

Glass, H. D. (1955). Fuel, Vol. 34, p. 253.

Golova, O. P., Krylova, R. G., and Nikilova, I. I. (1959). Vysokomol. Soedin, p. 1235.

Golova, O. P. (1975). Chemical reactions of cellulose under the action of heat, Usp. Khim, Vol. 44, p. 1454.

Graboski, M. S. (1981). "Properties of biomass relevant to gasification," Biomass Gasification: Principles and Technology, T. B. Reed, ed., Noyes Data Corp., Park Ridge, NJ.

Graham, R. G., Freel, B. A., Huffman, D. R., and Bergougnou, R. (1994). Applications of Thermal Processing of Biomass, A. V. Bridgwater, ed., Blackie Academic and Professional, London, England.

Gyulai, G., and Greenhow, E. J. (1974). J. Therm. Anal., Vol. 6, p. 279.

Halpern, Y., and Patai, S. (1969). Israel J. Chem., Vol. 7, p. 673.

Hemingway and Connors, eds. (1981). Adhesives from Renewable Resources, ACS Monograph 385, American Chemical Society, Washington, DC.

Heyns, K., and Klier, M. (1968). Carbohyd. Res., Vol. 6, p. 436.

Hoffman, I., Schnitzer, M., and Wright, J. R. (1959). Anal. Chem., Vol. 31, p. 440.

Hollings, H., and Cobb, J. W. (1923). Fuel, Vol. 2, p. 322.

Hopkins, M. W., DeJenga, C. I., and Antal, J. J., Jr. (1984). Solar Energy, Vol. 32, p. 547.

Hounimer, Y., and Patai, S. (1967). Tetrahedron Letters, Vol. 14, p. 1297.

Inman, R. (1981). "The potential biomass resource base," Biomass Gasification: Principles and Technology, T. B. Reed, ed., Noyes Data Corp., Park Ridge, NJ.

Institute of Gas Technology, Coal Conversion Systems Technical Data Book. (1978). DOE Contract EX-76-C-01-2286, Department of Energy, Washington, DC.

Jenkins, B. M., and Ebeling, J. M. (1985). Energy from Biomass and Waste IX, IGT, p. 371.

Johnson, J. L. (1979). Kinetics of Coal Gasification, John Wiley & Sons, New York, NY.

Kato, K., and Komorita, H. (1968). Agr. Biol. Chem., Tokyo, Vol. 32, p. 715.

Kilzer, F. J., and Broido, A., (1965). Speculation on the nature of cellulose pyrolysis, Pyrodynamics, Vol. 2, p. 151.

Kilzer, F. J. (1971). Cellulose and Cellulose Derivatives, Vol. 5, 2nd Ed., Wiley Interscience, New York, NY, p. 1015.

Kirk and Othmer. (1983). Carbohydrates, Encyclopedia of Chemical Technology, Vol. 1.

Kislitsyn, A. N. (1971). Zh. Prikl. Khim., Lenningrad, Vol. 44, p. 2587.

Kissinger, H. E. (1957). Anal. Chem., Vol. 29, p. 1702.

Klein, N. T., and Virk, P. S. (1981). Preprints of Div. Fuel Chem. Am. Chem. Soc., Vol. 26, p. 77.

Kobayashi, H. (1978). Ph.D. Thesis, Massachusetts Institute of Technology, Cambridge, Massachusetts.

Kujirai, T., and Akahira, T. (1925). Sci. Papers Inst. Phys. Chem. Res., Tokyo, Vol. 2, p. 223.

Lai, Y. Z., and Shafizadeh, F. (1974). Carbohyd. Res., Vol. 38, p. 177.

Laidler, K. J. (1969). Theories of Chemical Reaction Rates, McGraw-Hill, New York, NY, p. 52.

LeChatelier, H. (1887). Bull. Soc. Fr. Mineral Cristallogr., Vol. 10, p. 204, translated by Dr. L. E. Nesbitt, Southern Colorado State College, Pueblo, CO.

Lede, J., Diebold, J. P., Peacocke, G. V. C., and Piskroz, J. (1997). Developments in Thermochemical Biomass Conversion, A. V. Bridgwater and D. G. B. Boocock, eds., Vol. 1, p. 27.

Lede, J., et al. (1980). Revue Phys. Appl., Vol. 15, p. 545.

Lewin, M., and Basch, A. (1977). "Fire retardancy," The Encyclopedia of Polymer Science and Technology (suppl.), John Wiley & Sons, Inc., New York, NY, Vol. 2, p. 340.

Lewis, W. C. M. (1918). J. Chem. Soc. (London), Vol. 113, p. 471.

Lincoln, K. A. (1980). Proceedings of the Specialists' Workshop on Fast Pyrolysis of Biomass, Copper Mountain, CO, SERI/CP-622.

Lipska, A. E., and Parker, W. J. (1966). Kinetics of pyrolysis of cellulose in the temperature range 250–300C, J. Appl. Polym. Sci., Vol. 10, p. 1439.

Liptay, G. (1972). Atlas of Thermoanalytical Curves, Heyden and Sons Ltd., New York, NY.

Mackenzie, R. C. (1970). Differential Thermal Analysis, Vol. I, Academic Press, New York, NY.

Madorsky, S. L. (1958). J. Res. Nat. Bur. Stand., Vol. 60, p. 343.

Madorsky, S. L., Hart, V. E., and Straus, S. (1956). Res. Nat. Bur. Stand., Vol. 56, p. 343.

Madorsky, S. L. (1964). Thermal Degradation of Organic Polymers, Interscience, New York, NY.

Martin, S. (1965). "Diffusion controlled ignition of cellulose materials by intense radiant energy," 10th Int. Symp. on Combustion, The Combustion Institute, Pittsburgh, PA, p. 877.

Milne, T. A., and Soltes, M. N. (1981). "Fundamental pyrolysis studies," Annual Report for Fiscal Year 1981, SERI/PR-234-1454.

Milne, T. (1981). Biomass Gasification: Principles and Technology. T. Reed, ed., Noyes Data Corp., Park Ridge, NJ.

Milne, T. (1981). "Pyrolysis—the thermal behavior of biomass below 600°C," Biomass Gasification: Principles and Technology, T. B. Reed, ed., Noyes Data Corp., Park Ridge, NJ, p. 91.

Mok, W., and Antal, M. J. (1981) "Effects of pressure on biomass pyrolysis and gasification," 13th Biomass Thermochemical Conversion Contracts Meeting, Arlington, VA.

Murphy, C. D. (1958). Anal. Chem., Vol. 30, p. 967.

Murty, K. A. (1972). Combustion Flame, Vol. 18, p. 75.

National Renewable Energy Laboratory (1994). "Biomass pyrolysis oil, properties and combustion," Conf. Proc. NREL/CP-433-7265.

Norton, F. H. (1939). J. Am. Ceramic Soc., Vol. 22, p. 54.

Patai, S., and Halpern, Y. (1970). Israel J. Chem., Vol. 8, p. 655.

Paucault, A., and Sauret, G. (1958). Compt. Rend. Acad. Sci., Vol. 246, p. 608.

Puddington, I. A. (1948). Can. J. Res., Vol. B26, p. 415.

Pyle and Zorar. (1984). Models for the Low Temperature Pyrolysis of Wood Particles, A. V. Bridgewater, ed., Butterworths, London, England, p. 201.

Ramachandran, V. S., and Bhattacharya, S. K. (1954). J. Sci. Ind. Res., India, Vol. 13A, p. 365.

Reed, T. B. (1993). "An overview of the current status of biodiesel," Proc. 1st Biomass Conference of the Americas.

Reed, T. B. (1981). Biomass Gasification: Principles and Technology, Noyes Data Corp., Park Ridge, NJ.

Reed, T. B. (1985). "Biomass gasification," Advances in Solar Energy, K. W. Boer and J. A. Duffie, eds., American Solar Energy Society, Boulder, CO, Vol. 1, p. 158.

Reed, T. B., Graboski, M. S., and Gaur, S. (1992). Biomass and Bioenergy, Pergamon Press, London, England, Vol. 3, No. 2, p. 111.

Reed, T. B., Mobeck, W. L., and Gaur, S. (1994). "Advances in thermochemical conversion, A. V. Bridgwater, ed., Blackie Academic and Professional, London, England, Vol. 2, p. 1214.

Reed, T. B. (1985). "Principles and technology of biomass gasification," Advances in Solar Energy, Vol. 2, K. W. Boer and J. A. Duffies, eds., Plenum Press, New York, NY.

Reed, T. B. (1981). A Survey of Biomass Gasification. Noyes Data Corp., Park Ridge, NY.

Richard, J. R., and Antal, M. J. (1994). Advances in Thermochemical Conversion, A. V. Bridgwater, ed., Blackie Academic and Professional, London, England, Vol. 2, p. 784.

Richards, G. N., and Shafizadeh, F. (1978). Aust. J. Chem., Vol. 31, p. 1825.

Risser, P. G. (1981). Biomass Conversion Processes for Energy and Fuels, S. S. Soffer and O. R. Zaborsky, eds., Plenum Press, New York, NY, p. 25.

Roberts-Austen, W. C. (1899). Proc. Inst. Mech. Engrs., Vol. 1, p. 35.

Rossi, A. (1984). Progress in Biomass Conversion, D. A. Tillman and E. C. Jahn, eds., Academic Press, New York, NY, Vol. 5, p. 69.

Sanderman, W., and Augustine, H. (1963). Holzroh U-werkstoff, Vol. 21, p. 305.

Satava, V., and Skvara, F. (1969). J. Am. Chem. Soc., Vol. 52, No. 11, p. 591.

Satterfield, C. N., and Sherwood, T. K. (1963). Role of Catalysis in Diffusion, Addison-Wesley, Reading, MA.

Schulten, H. R., Bahr, U., and Gortz, W. (1981). J. Anal. Appl. Pyrolysis, Vol. 3, p. 137.

Schwenker, R. F., and Beck, L. R. (1960). Textile Res. J., Vol. 30, p. 624.

Shafizadeh, F., and Bradbury, A. G. W. (1979). J. Appl. Polym. Sci., Vol. 23, p. 1431.

Shafizadeh, F., and Lai, Y. Z. (1973). Carbohyd. Res., Vol. 31, p. 57.

Shafizadeh, F. (1980). Proc. ASME Annu. Mtg., Vol. 31, p. 122.

Shafizadeh, F., Furneaux, R. H., Cochran, T. G., Scholl, J. P., and Sakai, Y. (1979). J. Appl. Polym. Sci., Vol. 23, p. 3525.

Shafizadeh, F. (1975). Industrial pyrolysis of cellulosic materials, J. Appl. Polym. Sci., Symp. No. 28, p. 153.

Shafizadeh, F., Sarkanen, K. V., and Tillman, D. A., eds. (1976). Thermal Uses and Properties of Carbohydrates and Lignins, Academic Press, New York, NY.

Soltes, E. J., and Milne, T. A. (1988). Pyrolysis Oils from Biomass, ACS Symposium Series 376, American Chemical Society, Washington, DC.

Stamm, A. J. (1956). Thermal degradation of wood and cellulose, Ind. Eng. Chem., Vol. 48, p. 413.

Stevens, D. J. (1994). Review and Analysis of the 1980–1989 Biomass Thermochemical Conversion Program, NREL Subcontract AAE-3-13029-01, National Renewable Energy Laboratory, Golden, CO.

Swern, D. (1979). Bailey's Industrial Oil and Fat Products, 2 Volumes, 4th Ed., John Wiley & Sons, NY.

Tang, W. K., and Eickner, H. W. (1968). USFS Forest Products Laboratory Research Paper, FPL-82, United States Forestry Service, Madison, WI.

Tateno, J. (1966). Trans. Faraday Soc., Vol. 62, p. 1885.

Tillman, D. A. (1978). Wood As An Energy Resource, Academic Press, New York, NY.

van Krevelen, D. W., van Heerdan, C., and Huntjens, F. J. (1951). Fuel, Vol. 30, p. 253.

Varhegyi, G., Szabo, P., and Antal, M. J. (1993). ACS National Symposium, Cellulose and Textile Div., Abstract No. 12, American Chemical Society, Washington, DC.

Varhegyi, G., Jakab, E., and Antal, M. J., Jr. (1995). Is the Broido-Shafizadeh Model for cellulose pyrolysis true? Energy and Fuels.

Varhegyi, G., Szekely, T., Till, F., Jakab, E., and Szabo, P. (1988). J. Therm. Anal., Vol. 33, p. 87.

Varhegyi, G. (1978). Thermochimica Acta, Vol. 25, p. 201.

Walker, J. R. (1970). The pyrolysis and ignition of cellulosic materials: a literature review, J. Fire Flamm., Vol. 1, p. 12.

Walker, P. L., Rusinko, F., Jr., and Austin, L. G. (1959). Adv. Catalysis, Vol. XI, p. 133.

Waller, R. C., Bass, K. C., and Roseveare, W. E. (1948). Ind. Eng. Chem., Vol. 40, p. 138.

Weber, D. R. (1982). Energy Information Guide, Vol. 1: General and Alternative Information Sources, ABC-Clio, Santa Barbara, CA.

Weinstein, M., and Broido, A. (1970). Pyrolysis-crystalinity relationships in cellulose, Comb. Sci. and Tech., Vol. 1, p. 287.

Wendlandt, W. W. (1961). J. Chem. Educ., Vol. 38, p. 571.

Wendlandt, W. W. (1974). Thermal Methods of Analysis, 2nd Ed., John Wiley & Sons, New York, NY.

Wiegerink, J. G. (1940). J. Res. Natl. Bur. Stand., Vol. 25, p. 435.

Williams, P. T., and Besler, S. (1994). Advances in Thermochemical Biomass Conversion, A. V. Bridgewater, ed., Blackie Academic and Professional, London, England, Vol. 2, p. 771.

Wilty, J. R., Wicks, C. E., and Wilson, R. E. (1984). Fundamentals of Momentum, Heat and Mass Transfer, 3rd Ed., John Wiley & Sons, New York, NY, p. 299.

Wittels, M. (1951). Amer. Mineralogist, Vol. 36, p. 615.

Wunderlich, B., and Bopp, R. C. (1974). J. Therm. Anal., Vol. 6, p. 335.

Wunderlich, B. (1990). Thermal Analysis, Academic Press, New York, NY.

Zsako, J. (1970). J. Therm. Anal., Vol. 2, p. 145.

Zsako, J. (1973). J. Therm. Anal., Vol. 5, p. 239.

# INDEX